WATER FOR NEW YORK

A Study in State Administration of Water Resources

WATER
FOR NEW YORK

A Study in
State Administration of Water Resources

Roscoe C. Martin

SYRACUSE UNIVERSITY PRESS

Library of Congress Catalog Card: 60-9946
© 1960 by Syracuse University Press

Manufactured in the United States of America

Preface

As MANY as a hundred individuals contributed time and energy in varying amounts to the research which went into the preparation of this volume, and at least fifty more read and commented on parts of the manuscript. A list of all such contributors would be quite long and cumbersome. Moreover, it might be taken to ascribe some measure of responsibility-by-association, which at least some who helped with the book would not wish. It seems better, then, to leave unnamed most of those who rendered assistance along the way, though with this general acknowledgment in addition to the personal thanks already tendered.

William M. Shear served as my research assistant throughout the research phase of the study. In addition, the following acted as special research assistants as noted: Charles W. Eliot for Chapter VII, Robert M. Gidez for Chapter VIII, and James Wilson for Chapter VI. My debt to my research assistants is best understood by the individuals involved, but it is great.

A number of persons undertook to read and criticize the manuscript in detail. Included among these were, first of all, several members of the staff of Resources for the Future, notably Edward A. Ackerman, Irving K. Fox, and Henry Jarrett. In addition, the following read and made many useful comments on the study: Henry C. Hart, Associate Professor of Political Science, University of Wisconsin; Charles McKinley, Professor of Political Science, Reed College; Norman A. Stoll, Attorney, Portland, Oregon (formerly General Counsel to the Bonneville Power Administration); Thorndike Saville, sometime Dean of the College of Engineering, New York University; and Harold G. Wilm, formerly Associate Dean of the New York State College of Forestry, now Conservation Commissioner of the State of New York.

Professors Charles M. Haar of the Harvard University Law School and Vincent Ostrom of the University of California (Los Angeles) gave Chapter IV their special attention. Those named contributed greatly to the improvement of the manuscript in revision; it is hardly necessary to emphasize that they are not to be held responsible for any shortcomings the book may be found to have.

Resources for the Future, Inc., played an important part in the making of this book, first through a contract with the author for support at the research stage, second through a generous contribution toward its publication. It goes without saying that RFF accepts no responsibility for the contents of the study. At the same time, both the author and the director of the Syracuse University Press wish to acknowledge their indebtedness for the assistance which made it possible to bring the book to fruition.

Every research undertaking has its cut-off date. That date for the present study was January 1, 1959. This means that the book takes no account of the developments of 1959, some of which hold promise of improvement in the administration of New York's water resources. Whether or not this promise will be realized is a matter for future determination. That of course is another and different story.

Roscoe C. Martin
Syracuse University

Spring, 1960

Contents

Maps

Tables

Introduction

THE importance of the study of water resources will hardly require emphasis for anyone who may have the interest and energy to pick up this book, for such a one may be expected to be generally conversant with the current trends which stamp water as one of the nation's major problems. That total use of water has more than doubled every twenty-five years since 1900; that dramatic new demands have sent consumption soaring to heights undreamed of a few short years ago; that water quality tends to deteriorate as need presses on supply; that the eastern states, long accustomed to consider their reserves adequate to any need, have been slow to take vigorous action but in recent years have revealed an increasing sense of responsibility in the water-resource field—these and a host of kindred facts lie in the public domain and do not require reiteration here. The significance of water resources as a subject for study does not require documentation.

Nor does the increasingly important place of public action in the management of water resources. Federal, state, and local governments have been active in that field since the beginning of the nation; and as in many another area, public activities there have grown rapidly of recent years. As in many another area, too, the practice of federalism, and some say the fundamental character of the federal system as well, has undergone significant change with the passing of time. The nature of government's role, therefore, gives pause for thought.

In a literal sense, the federal system concerns the distribution of powers between the states and the national government. The Constitution emphasized the pivotal position of the states by vesting in them the residual powers of government, at the same time granting to the national government certain important enumerated powers. Among the latter were the powers to regulate commerce, levy taxes,

1

and make expenditures for the general welfare. It was recognized from the beginning that the practice of American federalism would take form in the interpretations given these and like constitutional clauses. The agencies of interpretation were several, but the basic process came to be one of congressional assertion through statute and judicial ratification or denial through court decision. This process, with generous (frequently determining) contributions by the executive branch, has brought the federal system from an uncertain borning in the Constitution to the complex but workable (or at any rate working) pattern of nation-state relations we know today.

As might have been foretold of a confessedly experimental institution, the pendulum of federalism has swung from state to nation to state to nation, though almost never quite so far back toward state sovereignty and almost always a little farther along the road toward national ascendancy. The last twenty-five years have witnessed a strong tide running toward the Federal government, with the Supreme Court going along on broad interpretations of the basic national powers, particularly, in the present context, those relating to commerce, taxes, and the general welfare. The result has been what some have called a "constitutional revolution," engineered by President and Congress and ratified by the Court. In the process of adjustment, the national government has greatly expanded familiar Federal activities in the water-resources field and has entered upon a number of major new undertakings. Federal water programs now include water power, flood control, pollution control, land management and soil conservation, small watershed protection, and, of course and always, navigation. The state is not precluded from substantial action in any of these fields, but what it is free to do in any particular case is conditioned (but not necessarily determined) by what the national government has already done, and by financial considerations. The comparative size of Federal and state purses is an ever-present factor which thus far has argued persuasively for a relax-and-enjoy-it attitude on the part of the state.

The upshot of recent developments is that the initiative in the area of water-resources administration has passed (or is rapidly passing) from the states to the nation; for it is the national government for the most part which conceives and implements the exciting new programs, and it is preponderantly the national government which pays the bill. It is not clear that this must continue to be so, but for the moment the states have been relegated by events to the role of junior partner in the fast-developing field of water management. The reserved powers of the states gradually have become left-over powers instead.

There is yet another aspect of federalism which is not usually identified as such and which receives a good deal less attention than the
pervasive problem of nation-state relations. From the point of view of
the states, federalism involves relations not only upward, which is traditional, but downward as well; for if the states are caught in the movement toward centralization, they are also, and by a strange inversion,
victims of an internal centrifugal force which threatens them from
below. What matters to the states is not alone constitutional definition,
which identifies them as the resting place of the residual powers of
government, but quite as importantly political and administrative
practice, which places them in an awkward position between the upper
millstone of national ascendancy and the nether one of local autonomy.
The active new channels which have opened up between the Federal
government and the local units in the last quarter-century have accentuated this problem.

The relation of the state to its local units is clear enough, in point
of law: it is the source of all local powers, and legally (and as a general
rule) may create and abolish local governments at will. Here if nowhere
else the state is sovereign—in law. In practice, no such thing is true,
for it must recognize the prerogatives and powers and even more the
traditions of local governments, as a matter of constitutional requirement
in some instances but more than that as a matter of political necessity.
The state cannot afford to exercise anything like all the legal power
it has vis-à-vis its local units. Further, as a practical consideration,
the state is dependent upon the active cooperation of the local units in
carrying many of its programs into effect. It treads softly, then, with
regard to the local governments, resigning to them much program initiative and devolving upon them a considerable part of the responsibility for the execution of state programs.

An objective appraisal of the trends in the vertical movement of
power and influence leads inevitably to the conclusion that the states
have lost ground as partners in the federal system.[1] The causes are
many and complex, but it may be hazarded that the states themselves
are not without blame for their downgrading. This observation leads
to contemplation of the nature of state responsibility and the manner
of its discharge. Disregarding such overwhelming developments as the
continuing technological revolution and America's changing role in international affairs, which it may be assumed state action could not have

---

[1] A careful recent analysis of these trends, their causes and their effects, is
afforded by The Commission on Intergovernmental Relations in its *Report to the
President for Transmittal to the Congress* (Washington, 1955).

modified materially, there remains an impressive area in which the states might have played a more significant role than they have. The cords which have bound the hapless Gulliver are four. The first is found in a constitution which almost uniformly inhibits rather than facilitates bold and vigorous action. A second arises from a dearth of dynamic leadership. A third lies in the reluctance of the states to expand their geographical jurisdiction by joint action as public problems increase in reach.

A fourth and final reason for state action which is frequently inept and ineffectual inheres in the administrative system. Administration occurs when officials and employees of the state undertake to carry into effect programs which have been agreed upon; associated primarily with the executive branch, administration has been characterized as that phase of the government cycle which represents action. Virtually all state employees except those attached directly to the legislature and the courts are engaged in administrative activity in some form. Further, the great bulk of state expenditures goes to administration, or to programs implemented by administrative action. The argument can be made, therefore, that state government consists primarily in administration; and the case is particularly strong with regard to government as the citizen sees it, for his normal contacts with government center on his relations with and observation of administrative personnel.

Considering the pervasiveness of administration and its significance for the success of governmental activities, it is strange both that its importance was so recently recognized and that even now it commands so frequently only incidental and sporadic attention. The problems of administration center on program planning, the structure of the administrative machine, staffing the organization, leadership and direction, the coordination of activities, and fiscal management and control. Upon their solution or, since they will never be permanently solved, their amelioration, depends the tone of the state's government and the effectiveness of its programs. That the states have lost ground relatively, that they have failed withal to discharge satisfactorily the responsibilities vested in them, is attributable in no small part to their neglect of administration. A fundamental test of the efficacy of a government is found in the answer to the question whether it does well what it sets out to do. The answer will be determined very largely by the tone and temper of the administrative organism.

These observations have special relevance to an important programmatic field, the management of water resources, which currently calls for attention in New York. The State has recognized the problem of water quantity and quality since Colonial days, and for upward of half a century has maintained an increasingly intricate organization to administer its expanding programs in the field. The basic issues which confront New York, as it moves into a period in which water problems will become more and more complex, are two. The first concerns policy: what is the proper area for state action as regards its water resources, and what direction should such action take in terms of program? The second involves administration: what organizational forms would best facilitate achievement of the programmatic goals sought? The seeming simplicity of these questions will not obscure their true nature, for they relate to the fundamental issues of government and are in fact as complex as any that might be asked. Their resolution is further complicated by the nature of the water resource, which is not neatly packaged for separate consideration but on the contrary ramifies throughout the whole of government. It is easy to call for the integration of water-resource activities and their administration; it is a quite different and much more difficult thing to marshal administrative energies in a way that will center attention on water resources without serious detriment to important related programs. Yet the challenge to thought and action is clear, for New York's water and water-related programs and their administration palpably leave much to be desired.

The present study raises some fundamental questions about the State's administration of its water resources. In so doing it accepts as valid four major propositions. The first, that man increasingly acquires the power to modify his environment, and that he now possesses in substantial measure the tools and the technical knowledge necessary to enable him to control that part of nature manifest in the water resource. The second, that public action is fundamental in a water-resource program, and that the state, and New York in particular, has an important role to play in the management of water resources. The third, that administration lies at the core of successful governmental action. The fourth, that government is susceptible of improvement by conscious action. Alexander Hamilton, addressing the people of New York in the first of the Federalist papers, written in 1787, raised the question ". . . whether societies of men are really capable or not of establishing good government from reflection and choice, ..." This is a fair question today. Hamilton answered it affirmatively. This study rests upon the conviction that he was right.

Part I: Setting

CHAPTER I

New York's Water Resources

WATER has long been known to move over the face of the earth in a systematic fashion. New York's first administrative agency established to consider the problems of water as such, the Water Storage Commission, in 1903 described the movement of water in these words:

> Like the circulation of the blood in the animal economy, water is raised from the ocean in the form of vapor by the force of the sun, and is diffused over the earth and condensed by atmospheric currents energized by the same central force, and after performing its vital function of stimulating life and growth in the forest and the field, it gradually returns again by its venous system of rills, brooks, creeks and rivers to the ocean. This return of the rainfall to the ocean is continuous, but exceedingly irregular.

A more matter-of-fact description of the process may be found in *Water*, the 1955 yearbook of the United States Department of Agriculture.[1] The movement of water "from ocean to sky to land to ocean" has come to be called the hydrologic cycle. The planned modification of this cycle may be termed water management.[2]

The character of the water management problem is determined by the

---

[1] William C. Ackermann, E. A. Colman, and Harold O. Ogrosky, "From Ocean to Sky to Land to Ocean," *Water* (Washington: U.S. Department of Agriculture, 1955), pp. 41-51. The whole of the section titled "Where We Get Our Water" is relevant to this subject.

[2] As employed in this study, "water management" or "management of water resources" relates primarily to the management of water in streams or rivers or collected in bodies such as lakes and reservoirs. The management of land to manipulate water resources, though recognized as a field of broad importance to water-resource administration, is a subject whose treatment would have opened up a sphere of inquiry considered beyond the scope of the present study. It is therefore omitted from consideration here.

kind, amount, and distribution (both in space and in time) of the water to be managed, and by the use demands made upon it. A study of water administration therefore begins with an appraisal of supply and demand. Such an appraisal requires, first, the careful gauging of both the amount and the behavior of water in its natural state. It requires, second, an evaluation of man's relation to the water resource, particularly, in the present context, as regards uses and needs. The first part of the exercise will provide an idea of the amount of water to be had, the second will afford an understanding of the demands to be satisfied. Both parts are wholly dependent upon the availability of accurate and reasonably complete records.

A Note on Water Records

On the supply side, it is doubtful that there is to be found anywhere in the country an accurate inventory of water resources for any considerable area.[3] In the first place, hydrology is a youthful science, its tools not yet sharp or precise. In the second place, economic, social, and political limitations make it difficult to take full advantage of what the hydrologists have to offer. In particular, hydrologic studies are costly and time-consuming, and their value is not generally recognized. In the third place, considerations of data quality apart, the administrative task of compiling and making currently available water resource data is an onerous and expensive one. Still, some progress toward a systematic inventory has been made, in New York as elsewhere. The Water Supply Commission as early as 1905 began a census of the State's water resources which succeeding agencies have continued to develop, with refinements, to this day. The State Planning Board in 1935 published a compendium of planning studies which contained a section on water, with maps indicating stream flow, rainfall stations, ground water supplies, critical watershed areas, and reservoirs built and proposed. The lengthy New England-New York Inter-Agency Committee (NENYIAC) report of 1955 contained a great deal of information on the amount and behavior of water in the State. Of these and like efforts it may be said that the results have been less than wholly satisfactory, principally for two reasons: first, the reports have been incomplete and sporadic; and second, too often they have rested upon secondary sources or upon superficial research.

[3] Harold E. Thomas has discussed the problems inherent in a water resources inventory in *The Conservation of Ground Water* (New York, 1951). See especially Chapter II.

A number of agencies currently are engaged in improving and expanding the inventory of New York's water resources. The Weather Bureau of the United States Department of Commerce collects data with regard to precipitation. The supplement for New York covering the years 1931-1952 employs records from 237 weather stations. Knowledge about precipitation, while by no means complete, is in a comparatively satisfactory state: as much is known about this phase of the hydrologic cycle as about any. A second task of inventory calls for a systematic recording of the flow of streams. This is done through some 210 gauging stations (1954 figure) scattered throughout the State. A considerable number of units and agencies, Federal and State and local, cooperate in this program, but primary responsibility rests on the United States Geological Survey on the one hand and the State Department of Public Works and the Water Power and Control Commission on the other. As in the case of precipitation, the body of information on stream flow is quite useful, though the data for both would profit greatly from more complete coverage and further refinement.

With respect to ground water, the information thus far developed is, for much of the State, quite elementary. Some twenty-five years ago the U.S. Geological Survey, with the cooperation of the Water Power and Control Commission, initiated a survey of ground water resources on Long Island, where such resources were (as indeed they are still) of the greatest importance. As the work on Long Island progressed the survey was extended to other parts of the State. By the end of 1955 seventeen reports, covering sixteen counties in full and four others in part, had been published. Investigations were under way in twelve other counties. In as many as twenty-six upstate counties virtually no work had been done. The ground water inventory is therefore far from complete. That New York's situation is not unique is indicated by Thomas' testimony that ". . . ground water conditions are known with reasonable reliability in only about 5 per cent of the nation's area. In three-fourths of the country there is practically no detailed information concerning the ground-water reservoirs." [4] A second major gap in technical data is found in the almost total lack of information, until quite recently, about evaporation and transpiration. Notwithstanding current scientific advances in the field, there appears to be general agreement among hydrologists that the tools for measuring evapo-transpiration are still so rudimentary as to necessitate caution in use of the results. The method long in use, that of substracting run-

[4] *Ibid.*, pp. 12-13.

off from precipitation, produces an answer which is at best a rude approximation.

Summarizing, existing knowledge about precipitation and stream flow in New York is gradually assuming useful form, that about ground water and evapo-transpiration remains in a pioneer state. It follows therefore that there has not been, nor in the circumstances which prevail can there be, a scientific inventory of New York's water resources. Even so, enough is known to make possible, if not a scientifically accurate stock-taking, at least some useful generalizations.

On the demand side the need is less for research than for systematic recording and reporting. The principal water users are readily identifiable, more especially since, in New York, the bulk of the consumers are either industries or public supply systems. It does not necessarily follow that all of these users know how much water they actually use, though presumably they do. In any case they could be encouraged, and if necessary required, by the State to keep records and make reports. Speaking of a study he had made, one of the country's foremost authorities on water demand recently expressed moderate satisfaction with his estimates of water used by public supplies, commenting at the same time that the use of water by industry was very difficult to estimate and that his figures for industry consequently were of limited value. Note the use of the word "estimate"; most figures for water use which cover any sizable area or any considerable number of users are estimates, and must be employed as such. The main conclusion to be drawn from these observations is that public agencies have not yet launched the sustained effort necessary to collect and reduce to manageable form data on water use. This comment applies specifically to the water-resource agencies of the State of New York. It is not meant to suggest that it would be easy to make the public records of water use either complete or wholly accurate, but there is no reason to doubt that they could be greatly improved with modest additional effort. In the meantime, it is necessary to employ water-use data gingerly, remembering that they rest for the most part on estimates and have little claim to statistical accuracy.

WATER SUPPLY

A New York statute offers this definition:

"Waters" or "waters of the state" shall be construed to include lakes, bays, sounds, ponds, impounding reservoirs, springs, wells, rivers, streams, creeks, estuaries, marshes, inlets, canals, the Atlantic Ocean within the territorial limits of the state of New York, and all other bodies of surface or underground water, natural or artificial, inland or coastal, fresh or salt, public or private (except those private waters which do not combine or effect a junction with natural surface or underground waters), which are wholly or partially within or bordering the state or within its jurisdiction.[5]

Embraced within this extremely broad concept are 70,000 miles of streams and 3,500,000 acres of inland lakes. The pattern of the rich and varied water resource begins to emerge from a study of Map 1, which outlines the State's principal watersheds. The 1918 Report of the Conservation Commission listed twenty major watersheds, the 1937 report of the Flood Control Commission six. The 1952 report of the Joint Legislative Committee on Natural Resources incorporated both concepts, concluding that New York has twenty-four watersheds, grouped into six major drainage basins. The accompanying map (1) shows fourteen watersheds. Seven watersheds drain toward the south, seven toward the north. In terms of area drained in New York, the Hudson is the largest by a wide margin, with the Susquehanna second in size; two (11 and 12) are not of much significance in the total picture. The headwater streams of five watersheds originate in the Adirondacks, which makes that range truly the "Father of Waters" for New York State. Outside the fourteen drainage basins are the Atlantic Ocean to the east and Lakes Erie and Ontario to the west, which already figure large in New York's total water resources and which are destined to be even more important in the future.

Nearly all fresh water comes from precipitation, primarily of rain and snow. New York, favored by winds from the Gulf of Mexico and the Atlantic Ocean, suffers no serious dearth of water from the sky.[6] Map 2 provides a study of the State's mean annual precipitation, which varies from 30 inches in parts of northwestern New York to 50 inches or more in the central Adirondacks and along the lower Hudson. Total annual precipitation for the State averages about 40 inches, approximately 84 per cent of which is attributable to rainfall, 16 per cent to snowfall. This amounts to an average of more than 2,000,000 gallons of water per year, or 5,480 gallons per day, for every person in the State. If the fall were uniformly distributed, each acre of the State's surface would receive on the average over 1,000,000 gallons of water a

[5] *McKinney's Consolidated Laws of New York, Annotated*, Book 44, *Public Health Law*, Article XII, Title 1, Section 1202 (b).

[6] See R. A. Mordoff, *The Climate of New York State* (Cornell Extension Bulletin 764, December, 1949), pp. 22-28, for a brief discussion of this subject.

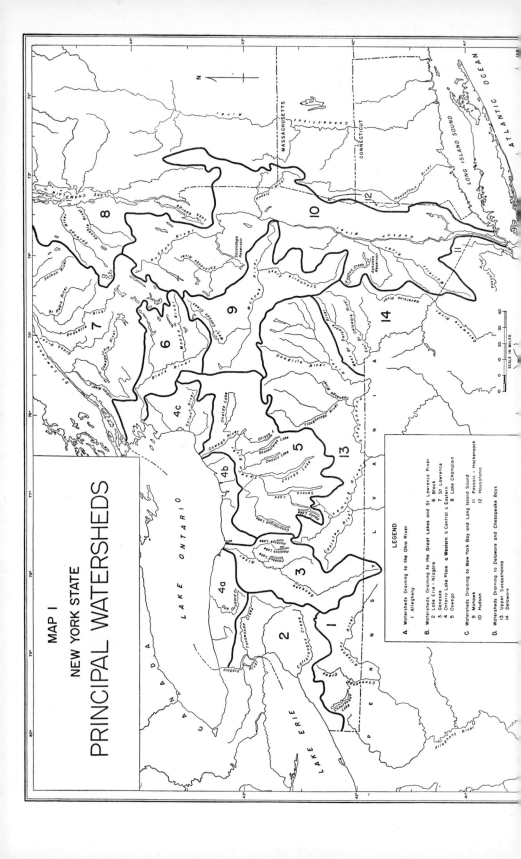

MAP I

NEW YORK STATE

PRINCIPAL WATERSHEDS

LEGEND

A. Watersheds Draining to the Ohio River
   1 Allegheny

B. Watersheds Draining to the Great Lakes and St Lawrence River
   2 Lake Erie - Niagara   6 Black
   3 Genesee   7 St Lawrence
   4 Ontario Lake Plain a Western b Central c Eastern
   5 Oswego   8 Lake Champlain

C. Watersheds Draining to New York Bay and Long Island Sound
   9 Mohawk   11 Passaic - Hackensack
   10 Hudson   12 Housatonic

D. Watersheds Draining to Delaware and Chesapeake Bays
   13 Upper Susquehanna
   14 Delaware

SCALE IN MILES

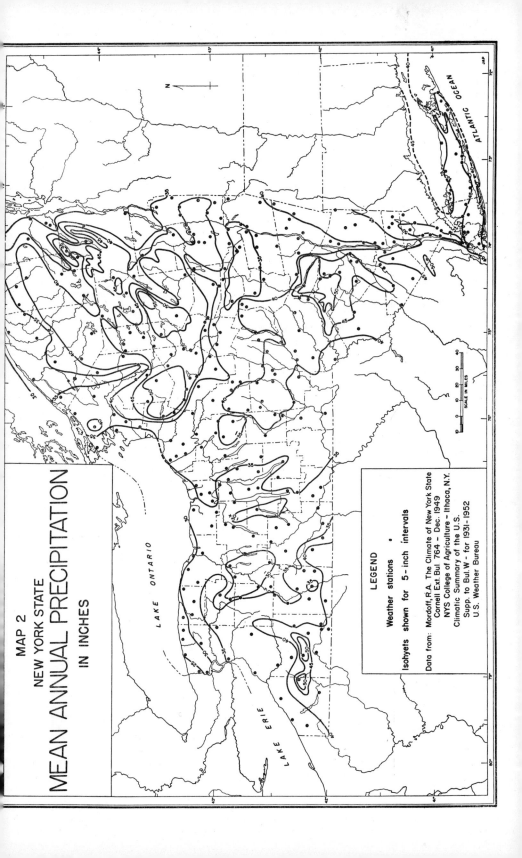

MAP 2

NEW YORK STATE

MEAN ANNUAL PRECIPITATION

IN INCHES

LEGEND

Weather stations •

Isohyets shown for 5 - inch intervals

Data from: Mordoff, R.A. The Climate of New York State
Cornell Ext. Bul 764 – Dec. 1949
NYS College of Agriculture - Ithaca, N.Y.
Climatic Summary of the U.S.
Supp. to Bul. W - for 1931-1952
U.S. Weather Bureau

LAKE ONTARIO

LAKE ERIE

ATLANTIC OCEAN

SCALE IN MILES

N

year from rainfall and snowfall, enough in the aggregate to fill a lake the size of New York City's new Cannonsville reservoir every weekday in the year. The map also indicates the distribution of the weather stations where precipitation is recorded.

One of three things happens to all the water which falls to the ground. First, as much as half (sometimes more) evaporates or is transpired by plant life into the atmosphere, and so is lost to productive use. Evaporation from water surfaces varies greatly, but nowhere in New York is it reported to be less than 15 inches in a year. One author has estimated that, of the average annual precipitation of 40 inches, approximately 19 evaporate and 11 are transpired by plants.[7] This estimate of 75 per cent loss in precipitated water through evapo-transpiration is high in comparison with the more cautious scientific calculations, which range from 50 to 60 per cent. If 50 per cent of the total precipitation be accepted as the loss figure through evapo-transpiration, the amount lost is about 1,000,000 gallons per year, or 2,740 gallons per day, per capita. Incidentally, contemplation of this subject suggests the vital importance of managing land and vegetation to minimize water losses by evaporation and transpiration.

Second, part of the water which reaches the earth through precipitation finds its way into streams. The proportion which runs off varies considerably in reaction to such local factors as character of the surface drained, topography, amount and kind of vegetative cover, and amount of moisture in the soil at the time of precipitation. It has been estimated that, for all New York, run-off approximates 30 per cent of total precipitation, or an average of about 12 inches a year. This amounts to some 600,000 gallons of water per person, or 300,000 gallons per acre. The figure of course varies greatly from one section to another: in parts of the Adirondacks the annual run-off reaches a total of 43 inches, while on sandy Long Island it is a fraction of that amount. Map 3 affords a visual impression of New York's network of streams, and of the number and distribution of its lakes as well. The Hudson River system dominates the eastern part of the State and the Susquehanna the south central. A number of streams feed into the St. Lawrence from their origins in the Adirondacks. All parts of the State have important surface water supplies, but in the western and more especially the northwestern portions (Map 1, Watersheds 4a, 4b, and 4c) the streams are fewer and smaller than elsewhere.

Third, water which falls to the earth may find its way into under-

[7] Pete Fosburgh, "Rain" (Reprint No. 145 from the *New York State Conservationist*, April-May, 1947).

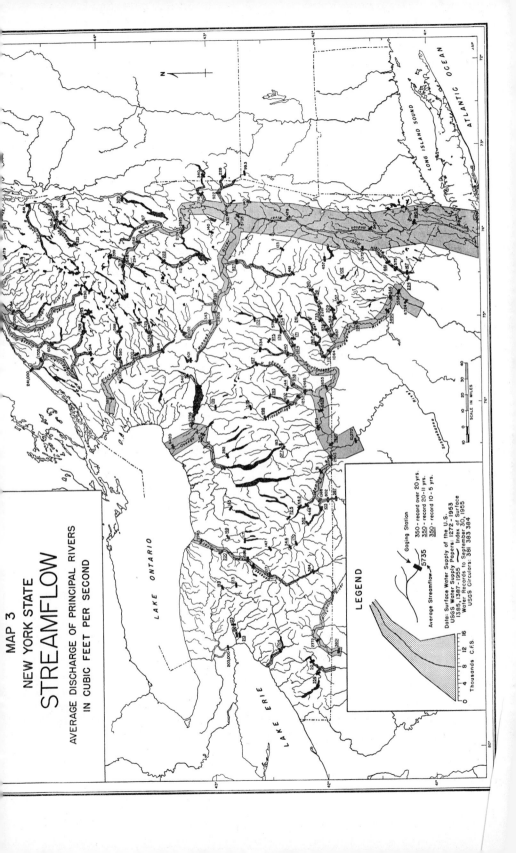

MAP 3

NEW YORK STATE

STREAMFLOW

AVERAGE DISCHARGE OF PRINCIPAL RIVERS
IN CUBIC FEET PER SECOND

LEGEND

■ Gaging Station

5735 ← Average Streamflow

350 - record over 20 yrs.
350 - record 20-11 yrs.
350 - record 10- 5 yrs.

Data: Surface Water Supply of the U.S.
USGS Water Supply Papers: 1272-1953
1385, 1387-1955 — Index of Surface
Water Records to September 30, 1955
USGS Circulars: 381 383 384

SCALE IN MILES
0   10   20   30   40

Thousands C.F.S.
0   4   8   12   16

LAKE ONTARIO

LAKE ERIE

ATLANTIC OCEAN

LONG ISLAND SOUND

ground streams and reservoirs as ground water. The proportion which penetrates the surface of the earth is substantially affected by the factors which influence run-off; indeed there is an important reciprocal relationship between ground water and stream flow. One estimate is that, over the whole State, an average of eight inches a year (20 per cent of the total average annual precipitation, 400,000 gallons of water per person, 200,000 gallons per acre) seeps into the soil as ground water. The extremes in ground water receptivity are found on Long Island and in the northwestern part of the State. On Long Island, sub-surface conditions are highly favorable to the storage of ground water, and large quantities of water for domestic, agricultural, and industrial uses are obtained from wells there. In upstate New York, on the other hand, the underground water-bearing deposits are much less hospitable to water storage; there are comparatively few extensive sand and gravel deposits, while relatively impermeable rock formations are fairly prevalent. By comparison with Long Island, therefore, upstate New York's ground water resources are not plentiful. Even so, ground water is widely distributed throughout the State, though apart from Long Island comparatively limited in quantity and frequently somewhat lacking in quality as well.

Palpably New York has an abundant supply of water. Whether, as is popularly believed, it has "enough" water depends on a number of factors. And whether, assuming an ample supply for the present, the resources will prove adequate to future requirements will depend on several developments, among them growth in use. This brings us to consideration of water demand.

WATER DEMAND

Water may be classified according to use several different ways. One classification turns on the effect of use on quantity. On the one hand, the water may be actually consumed, either through evaporation or through absorption into a product, and so lost to further use. On the other it may be, so to speak, borrowed and returned to the common supply after use substantially unchanged in quantity, if sometimes drastically modified in quality. Irrigation is the principal consumptive user of water, though even there a high percentage of the water used frequently finds its way back into a stream or into the soil. The other major users do not actually consume any considerable part of their total demand; their uses are primarily nonconsumptive. It is estimated that, apart from irrigation, upward of 90 per cent of all water used is returned either to the streams or to the ground. Inasmuch as only a small part of the water used in New York goes to irrigation, it is likely that not more than 5 per cent of all water used here is consumed. This means that most water is available for re-use, and some water in fact is used not only for a number of different purposes but a number of times as well.

A more useful classification for present purposes is that offered by the withdrawal/nonwithdrawal concept.[8] The basis for classification here depends on whether the water is actually withdrawn from its source or is utilized without withdrawal. The principal nonwithdrawal uses are for navigation, waste disposal, recreation, and the conservation of fish and wildlife. These uses cannot be measured in terms of the amount of water used, although, as MacKichan points out, they are highly significant not only in their social and economic effects but also in their consequences for other water uses.[9] The major withdrawal uses of water are for water power plants, industries, public supply systems, rural homes and farm establishments, and irrigation. They can be measured, or more accurately, in view of the state of the records, estimated. A number of hydrologists have sought during the last ten years to determine the amount of water used, prominent among them MacKichan (cited above) and Picton.[10] The work of these two provide the principal base for the present summary.

Table 1 provides both an estimate of the withdrawal uses of water in New York in 1955 and trend estimates for the period 1950-1955. It is evident at once that water power is overwhelmingly the largest user of water. Most of this water is used for the generation of electricity. Very little of it is consumed in the process, virtually all being returned to the streams or to the ground. Of the other four users listed, self-supplied industries use more than three times as much water as the remaining three combined. The chief industrial uses are for condenser cooling purposes and for air conditioning. Public (largely municipal) supply systems provide water for domestic use, street cleaning, fire protection, the watering of lawns and gardens, and industries. Rural uses, defined to cover homes and farm establishments not drawing upon

[8] Kenneth A. MacKichan employs this concept in his studies estimating the use of water ("Estimated Use of Water in the United States, 1950," Geological Survey Circular 115, May, 1951; and "Estimated Use of Water in the United States, 1955," Geological Survey Circular 398, 1957).

[9] "Estimated Use of Water in the United States, 1955," p. 12.

[10] Walter L. Picton, "Summary of Information on Water Use in the United States, 1900-1975" (U.S. Department of Commerce, Business Service Bulletin 136, January 1956).

TABLE 1

Changes in Withdrawal Uses of Water in New York
1950-1955

Use	Water Used (million gallons per day)		Per Cent Change
	1950*	1955†	
Water power	116,000	130,000	+12
Self-supplied industrial	5,100	6,757	+32.5
Public supplies	1,890	1,960	+3.7
Rural	170	120	−29
Irrigation	28	47	+70
Total, other than water power	7,188	8,884	+23.6

* Adapted from Kenneth A. MacKichan, "Estimated Use of Water in the United States, 1950" (Geological Survey Circular 115, May, 1951), Table 1, pp. 6-7, as modified by a letter, MacKichan to author, dated January 22, 1959.

† From Kenneth A. MacKichan, "Estimated Use of Water in the United States, 1955" (Geological Survey Circular 398, 1957), Table 9, p. 13.

public water-supply systems, constitute a relatively minor demand in New York, as does irrigation.

More significant than the raw data for any particular year are the trends. Table 1 indicates that, from 1950 to 1955, water use increased considerably in all categories save one. Use for water power increased 12 per cent, while use by self-supplied industries grew by almost one-third. The growth in use by public supply systems was somewhat less than 4 per cent, which suggests that such systems are reasonably mature and therefore are not likely to experience rapid growth in the future. Rural uses decreased by 29 per cent, which probably indicates increasing rural reliance on public supplies. The most spectacular development was in irrigation use, which increased 70 per cent. Agriculture still uses comparatively little water in New York, but the growth of irrigation nevertheless is significant. Over-all, and excluding water power (which is so overwhelming in amount used that it would dominate the trend), the State increased almost 24 per cent in gross water use from 1950 to 1955. During the same period, New York's population growth was 5.6 per cent. Water use thus grew more than four times as fast as population. Another significant index of growth is found in the gallons-per-day-per-capita figure, which increased 14 per cent during the same period.

The long-range prospect is for continued steady growth. Projections for New York are lacking, but Picton has made an interesting trend study for the country as a whole.[11] His analysis shows that total water use increased 94 per cent from 1940 to 1955. He forecasts a total use increase of 72.5 per cent from 1955 to 1975, with the several use categories expanding as follows: self-supplied industries, 106 per cent; public supplies, 75 per cent; rural uses, 33 per cent; and irrigation, 42 per cent. The projected increase in use in gallons per day per capita during the twenty-year period is 34 per cent. If New York follows the projected national trend, its total use (excluding water power) in 1975 will be 15,374 million gallons per day.

New York uses a great deal of water, and there is every likelihood that it will use much more in the future, for both gross and per capita figures are increasing steadily. Still there is no prospect of an over-all shortage. For the country as a whole, 1955 estimates varied from 1,458 (MacKichan) to 1,598 (Picton) gallons per day per capita for all water uses. New York's average annual precipitation provides 2,740 gallons per day per capita after allowance is made for loss through evapo-transpiration. More water runs off (1,644 gallons per day per capita) in the State's streams than is required for total water use, assuming New York's use and that of the country to be comparable. Further, an estimated 1,096 gallons per day per capita enter the earth as ground water, and this constitutes an added and very important water resource. There are, moreover, two additional facts which work in New York's favor. First, use estimates are cumulative, which means that every gallon of water is counted separately for each use made of it. Use figures therefore are gross, and relate not to individual (net) gallons but to the total amount used. This is especially significant in view of the fact that some water is used several times. Second, only a start has been made toward the conservation of supplies through such practices as the recirculation of water used by industry. The water used undoubtedly can be made to go much farther than it has gone in the past. These considerations add measurably to the State's already favorable position with regard to its water resources.

WATER RESOURCES AND ADMINISTRATION

This is not to suggest that New York, because of its generous resources, is immune from the necessities implied in the term "water management." The water resources are not spread evenly throughout

[11] *Ibid.* See especially Table 1, p. 4.

the State, nor is the supply uniform for any particular place through the four seasons. Moreover, the very prodigality of the resources in some sense creates an issue; for if water is plentiful over much of the State, the supply in other areas is insufficient by comparison. Clearly nature's system for distributing water is not an efficient one in terms of man's needs.

The problem of water distribution assumes two principal forms. First, oftentimes there is alternately too much and too little water at a particular place. Precipitation is notoriously erratic, with virtually all of the snowfall concentrated in the four winter months and with rainfall favoring the winter and spring seasons. A direct result of uneven precipitation is found in alternating wet and dry periods. An indirect consequence is reflected in variations in stream flow, which sometimes are very great. Most of the streams originating in the Adirondacks are subject to flash floods, as are the headwaters of the Delaware, the Susquehanna, and the Allegheny. Ratios of high flow to low flow of 25:1 are not uncommon, and much higher ratios have been recorded. The natural variations in precipitation and stream flow produce a grave problem of water distribution *in time;* the issue is not one of total amount available, but of availability when needed—and, of course, of protection against over-abundance. The problem is to mitigate nature's vagaries to the end that water may be at hand when man's needs require it.

A companion distribution problem concerns *place;* it arises from the fact that New York's water, though bountiful in amount, is not always available where it is needed. In the summer of 1949, to illustrate, New York City suffered a severe water shortage, not because there was insufficient water for all the State's people but because arrangements had not been completed to bring water to the City in sufficient amounts. A relative shortage of water persists on the Ontario Plain, notwithstanding adequate sources are to be found nearby. The problem in these and like instances is one not of inadequate supply but of logistics, of transporting the water from where it is to where it is needed.

The purposeful modification of nature's water distribution system in the interest of human needs is the primary task of water management. The task involves first the levelling of water supply peaks and the alleviation of depressions, and second the redistribution of water resources according to place of need. As a student of the subject recently put it, the first is a "from now to then" problem, the second concerns "from here to there." Both require the storage of surplus water, and the second requires its transport as well. Further, if the natural distribution system is to be modified over an appreciable area, the plans to that end must be made and executed on a large scale. The problem of water quantity, using the term in a broad sense, is the first large problem for which public administration must assume major responsibility.

Most water uses, we have noted, are not consumptive, but return the water used to its source substantially undiminished in quantity though often materially modified in quality. Quality change may result from industrial use which raises the water temperature or from the infusion of chemicals or industrial wastes. It may come about also through the deposit of municipal sewage, which is an important cause of pollution. The control of water quality, the keeping (or rendering) of water fit for use, is therefore a second major problem of administration.

Knowledge about the water resource is in an elementary state, as has been observed. Yawning gaps exist in basic data, records in important areas are fragmentary, and reports are spotty and incomplete. Research, records, and reports therefore emerge as a third major problem area requiring the attention of administration.

New York is well aware of these and like problems, and for upward of fifty years has maintained an administrative organization to cope with them. This organization, which has grown increasingly complex through the years, administers a number of varied and wide-ranging water programs in behalf of the State. The remaining chapters of this study undertake to show how New York has responded to man's impact on the water resource, and to appraise the successes and failures of the policies it has pursued.

CHAPTER II

Water Programs

NEW YORK'S many-sided water program may be discussed to advantage in terms of background, incentives to program development, and some general observations on program growth. The purposes here will be to discover how the several major programs took shape, what they consist of at the present time, and why they developed as they did. Attention will be focused on program activities rather than administrative organization and action, subjects reserved for consideration in later chapters.

PROGRAM DEVELOPMENT

It is not the purpose here to write a history of New York's water programs. Such an undertaking would require volumes; moreover, adequate histories of at least some water programs have been written— there are, for a single example, a number of books dealing with the history of the inland waterways system. The intention rather is to highlight the important steps taken in the development of each of the major programs, in order to provide perspective for an understanding of current program activities. The programs selected for attention concern navigation, water supply, water quality, water power, water control, a miscellaneous group of three activities of lesser scope, and two emergent developments which promise to become significant.

Navigation

One of the oldest uses to which mankind has put water is that of transportation. Men came from the Old World to the New by sea, and their travel on these shores in the early days was confined almost exclusively to natural waterways. It did not take the settlers long to learn that the territory that was to become New York was equipped with a wide-flung system of inland waterways. As early as 1724 the surveyor-general of the Colony referred to the need of river improvement for the facilitation of commerce, citing the cost of transporting furs and Indian supplies between the interior lakes and the Hudson River as the basic reason for his concern. In 1768 Sir Henry Moore, Governor of the Colony, recommended to the Legislature a plan for the improvement of inland navigation. In 1784 Christopher Colles, who ten years before had been commissioned to devise a water supply system for New York City, suggested construction of a system of canals; and two years later a member of the Assembly introduced a bill ". . . for improving the navigation of the Mohawk River, Wood Creek, and the Onondaga River, with a view of opening an inland navigation to Oswego, and for extending the same, if practicable, to Lake Erie. . . ." [1] This bill, the first to propose official state sponsorship of a canal west from Albany to Lake Erie, was not well received by the Legislature, which took no action on it.

The first tangible step toward canal development occurred when, in 1792, the Legislature passed "An Act for Establishing and Opening Lock Navigation within the State" (Chap. 40, Laws of 1792). Under the terms of this act a number of prominent citizens organized two private lock companies, one to open navigation between the Hudson River and Lake Champlain, the other to improve the waterway westward to Lake Ontario and Seneca Lake. The western company achieved some success, and for the twenty years following 1797, small boats travelled as far west as Seneca Lake along the general route later developed as the Erie Canal.

Still action by the companies did not produce satisfactory results, and talk of a canal across the State grew in volume in the early eighteen-hundreds. In 1808 a member of the Assembly from Onondaga County proposed to the House that a joint committee be appointed to consider the matter of a canal linking the Hudson and Lake Erie. The frugal representative of what was even then a cautious constituency had no idea that the State would assume any part of the financial burden that

---

[1] Thomas F. Gordon, *Gazetteer of the State of New York* (Philadelphia, 1836), in Chapter VI gives a very useful account of the early days of the canal movement. The report of the Erie Canal Centennial Commission titled *The Erie Canal Centennial Celebration, 1926* (Albany, 1928) offers a brief history of the canal system. Especially useful is an address by Senator Henry W. Hill, which appears on pp. 68-82.

might result from his bold suggestion; instead, he sought action to the end ". . . that Congress may be enabled to appropriate such sums as may be necessary to the accomplishment of that great object." Even so, his proposal was received ". . . with such expressions of surprise and ridicule, as are due, to a very wild, foolish project." [2]

The fortunes of the proposed canal system took a turn for the better when, in 1810, the Legislature by concurrent resolution agreed to the appointment of commissioners to examine the condition of western navigation, to make any survey deemed necessary, and to report at the next session of the Legislature on the subject of navigation between the Hudson and Lake Erie. The key member of this commission was Senator DeWitt Clinton, who was even then Mayor of New York City and who was destined to serve two terms (1817-1822 and 1825-1828) as Governor. Clinton brought his considerable influence and great enthusiasm to bear on the problem, and by 1815 he and his associates had generated widespread interest in the Hudson-Erie canal proposal. In that year a memorial drawn by Mr. Clinton commanded wide attention, and public meetings were held in behalf of the canal in several counties. As a result, petitions bearing thousands of signatures were presented to the Legislature. Clinton, thus fortified, brought his memorial before the Assembly early in 1816, and in response the Legislature passed an act appointing commissioners to make definite plans for canals to connect the Hudson with Lake Champlain to the north and Lake Erie to the west. So convincing was the report of the commissioners that, on their recommendation, the Legislature in 1817 passed an act authorizing the beginning of work on the Erie and the Champlain Canals (Chap. 262, Laws of 1817). The preamble to the act is interesting; it reads, in part:

> WHEREAS, navigable communication between Lakes Erie and Champlain, and the Atlantic Ocean, by means of canals connected with the Hudson River, will promote agriculture, manufactures and commerce, mitigate the calamities of war, and enhance the blessings of peace, consolidate the union, and advance the prosperity and elevate the character of the United States: and WHEREAS, it is the incumbent duty of the people of this State to avail themselves of the means which the Almighty has placed in their hands for the production of such signal, extensive and lasting benefits to the human race: *Now, therefore,* . . .

DeWitt Clinton, now Governor Clinton, enjoyed the pleasure of turning the first spade of earth for the Erie Canal on July 4, 1817. He

[2] Gordon, *op. cit.,* p. 71.

seized the occasion to deliver an address which was prophetic of the good days to come with the growth of commerce stimulated by the canal. He also indicated that he saw quite clearly the significance of the forthcoming development for New York City.

Work proceeded apace on both the Champlain and the Erie, although the latter commanded the more attention. The Erie Canal was completed in 1825, and on October 26 of that year the Seneca Chief, described as an "elegant packet," moved into the new canal from Lake Erie, to the accompaniment of a series of batteries which carried the news all the way to New York City and back again.[3] Of the Erie it was said, "New York has built the longest canal in the world in the least time, with the least experience for the least money and for the greatest public benefit." [4]

The canal system expanded rapidly in the years following 1825. The Delaware and Hudson Canal, authorized in 1823, was opened to traffic in 1827. From 1826 to 1835 five lateral canals were completed. Moreover, in the latter year work began on the first enlargement of the Erie, a project designed to deepen its channel from four to seven feet. The canal system was a success from the beginning. Starting with a modest total of 291,000 tons in 1824, the trend of traffic was substantially upward for many years. The glorious days of canal transport reached from 1860 to 1890; the big years were from 1868 to 1874, a seven-year period during which traffic averaged more than 6,000,000 tons a year. The biggest year of all time was 1872, when the canal system carried 6,673,370 tons. After 1890 commerce declined gradually but persistently, reaching its low point in 1918 when the system carried less gross tonnage than in any year since 1834. Thereafter, with the opening of the new and expanded Barge Canal, tonnage began to climb again until, following a setback during the second World War, it reached a postwar high of 5,211,472 in 1951.[5]

There were those who believed the canal system to be doomed as the last century drew to a close. In 1882, the people approved a constitutional amendment making the canals toll-free. The next year the state engineer, declaring the free canals a failure, concluded that it was ". . . foregone and inevitable . . . that the canals must go." [6] The

---

[3] Carl Carmer has described this event in prose fitting the occasion. See *The Hudson* (New York, 1939), p. 219.

[4] Erie Canal Centennial Commission, *op. cit.*, p. 9.

[5] Joseph H. Salmon, *Economic Survey of the New York State Barge Canal* (New York, no date), p. 81.

[6] Noble E. Whitford, *History of the Barge Canal* (Albany, 1921), p. 21.

general spirit of gloom, of which this was but an example, spurred action to modernize the canal system; and in 1903, following a report by a committee appointed by Governor Theodore Roosevelt, the Legislature passed an act providing for the building of a modern barge canal (Chap. 147, Laws of 1903). Construction of the new system was not completed until 1918.

In 1956 the New York State Barge Canal system, stripped of a number of uneconomic tributaries, totaled 524 miles in length. Supported by 57 locks, the system had a minimum depth of 12 feet and accommodated boats to 300 feet in length and 43.5 feet in beam.[7] All during the nineteenth century forest and agricultural products dominated the canal tonnage. Today its principal business is the transport of petroleum and grain, which in 1958 contributed 78 per cent of the total tonnage. Petroleum and its products alone accounted for 67 per cent.[8]

There is no doubt that the state canal system, which means basically the Erie Canal and its tributaries, abundantly achieved the purposes for which it was conceived. It opened up western New York not only to trade but also to settlement, afforded a cheap and direct link between the Great Lakes and the eastern seaboard, and contributed handsomely to the growth and prosperity of New York City. The Barge Canal has carried on in the tradition of the Erie, if in somewhat more businesslike and less romantic fashion. Apart from its major function as an artery of transportation, the canal system serves a number of other purposes. Its contribution to flood control throughout its large basin is important. It constitutes a drainage system for the whole of central New York. This is especially significant for the Montezuma swamp area, where the canal has made large tracts of rich bottom land available for cultivation. It affords a modest source for hydroelectric power. It provides important recreational facilities, particularly on such lakes as Seneca, Cayuga, and Oneida, whose levels are carefully controlled by canal operations. It serves as a source for domestic and farm water supply, and bids fair to become even more important in this respect as agriculture makes good its claim to water for irrigation.[9] All these are auxiliary but not insignificant services. They combine with navigation to make the Canal a continuing force in the economic and social life of the State. The significance of its role was recognized by Senators

[7] *Rules and Regulations Governing Navigation and Use of the New York State Canal System* (July 1, 1956), pp. 29-30.

[8] Department of Public Works, Division of Operation and Maintenance (Canals, Waterways and Flood Control), *Trade and Tonnage*, Season of 1958, p. 10.

[9] *New York State Commerce Review*, Vol. 7, No. 2 (February, 1953), pp. 12-13.

Irving Ives and Jacob Javits of New York, who on July 17, 1957, addressed a joint letter to the Senate Public Works Committee requesting that body to adopt a resolution authorizing the Corps of Engineers to make a study of the possibilities of major improvements in the canal system.[10]

Other state navigation activities include participation in an interstate compact which provides the legal basis for the Port of New York Authority; sponsorship of the Albany Port District, which was created by legislative act in 1925 for the purpose of developing the Port of Albany; a continuing interest in the Hudson River from Albany south, where the Corps of Engineers is currently engaged in deepening the channel from 27 to 32 feet; and a vigorous concern in the St. Lawrence Seaway. The ports of Albany and New York and the river channel connecting them are, of course, vitally related to the canal system, for all are parts in a total transportation system. Regarding the Seaway, the positions of the State and (more especially) New York City are ambivalent. On the one hand, there is a natural suspicion of any newcomer that might threaten the extensive and costly navigation system; while on the other some parts of the State undoubtedly will benefit from Seaway operations, and the City itself might profit materially if the proposed improvement of the Richelieu Canal (connecting Lake Champlain with the St. Lawrence) should go through. The ultimate effects of the Seaway on New York are by no means clear.

Water Supply

New York has plenty of water, or so common belief holds; wherefore the State was late in entering the field of water supply, with the exception of a very few special situations. The principal exception has been provided by New York City, which has had the cooperation of the State since before 1800 in its quest for new water sources. Apart from these few special cases, the State's activities in the water supply field have been limited to this century.

The first general law dealing with water supply as such was passed in 1905 (Chap. 723, Laws of 1905). Creating a State Water Supply Commission, it provided (in Sec. 2) that

No municipal corporation or other civil division of the state, and no board, commission or other body of or for any such municipal cor-

[10] *Syracuse Post-Standard*, July 18, 1957. There are those who question the economic soundness of the Barge Canal as a "business enterprise." No attempt has been made in this study to evaluate the economics of canal operations.

poration or other civil division of the state shall, after this act takes effect, have any power to acquire, take or condemn lands for any new or additional sources of water supply, until it has first submitted the maps and profiles therefor to said commission, as hereinafter provided, and until said commission shall have approved the same.

It will be noted that the act covered only public water supplies, and only new or additional sources. A loophole immediately appeared in that the law was interpreted to exclude from regulation private companies serving municipal corporations. This defect was remedied by an act passed the next year (Chap. 415, Laws of 1906) to bring such companies under the regulation of the Water Supply Commission as regards "new or additional sources of water supply."

The new commission found a virgin field awaiting exploration and development. Actually it was created simultaneously with and as a companion agency to the New York City Board of Water Supply, with which it immediately cooperated to complete arrangements for the City's Catskill water system.[11] In its second annual report, the Commission identified a number of problems—pollution abatement, water power regulation, river improvement, and flood control chief among them—which indicated recognition that the law of 1905 and the agency created by it were entering into pioneer territory.

The State's program with respect to water supply has been almost entirely facilitative, regulative, and supervisory. A number of significant trends have characterized the half-century since the passing of the first important act relating to water supply. First, the State has undertaken a continuing responsibility for the allocation of sources of supply. This it has done through the Water Supply Commission and its successors. "Primarily the function of the Commission (with respect to water supply) is the equitable distribution of the water-supply resources of the State among the various communities therein, a matter of great and constantly increasing importance."[12] This policy is implemented chiefly through administrative control over plans and procedures for any new sources of water supply, in harmony with the original act of 1905 (which remains in effect without important substantive change). The drilling of large new wells on Long Island, for example, has been placed under administrative control by legislative act.[13] That the problem of water supply has not been completely solved

[11] This aspect of the subject is developed at some length in Chapter V, below.

[12] Conservation Commission, *Sixth Annual Report* (1916), p. 45.

[13] Originally the law covered wells designed to produce 69 gallons of water per minute (approximately 100,000 gallons a day), but in 1954 the limit was reduced

is evidenced by the experience of the northwestern counties to the west of Rochester. There as long ago as 1912 the Conservation Commission laid out an imaginative plan which called for a reservoir on Little Tonawanda Creek (at Linden) and for a grid of pipelines serving some 60 cities and villages from Rochester west to the Niagara River and from Batavia north to Lake Ontario.[14] Forty-five years later the Northwestern New York Water Authority was still wrestling with the same problem. Another unsolved supply problem concerns irrigation, which is already feeling a pinch and which undoubtedly will require increased amounts of water in the years to come.

Second, the State has played an increasingly energetic role with regard to the assurance of water quality. This story will be treated briefly in the following section.

Third, the movement among the municipal and other local water supply systems (primary objects of state concern) has been in the direction of public ownership. "This tendency appears to be due to the low rate of profit earned by these companies, and to the difficulty experienced by all public service corporations in obtaining funds for renewals and extensions." [15] So far has this trend progressed from the time when virtually all water supply systems were privately owned that approximately 60 per cent of all public water supply systems in the State are now government-owned.

Fourth, whereas the laws in the past favored the multiplication of many small systems, the weakness of such systems is now recognized and the statutes currently favor fewer and larger systems. Contemporary policy manifests itself in the authorization of county-wide systems, inter-governmental arrangements, and "union" districts designed to pool the resources (and the needs) of rural and suburban users.[16]

In 1954 there were 1,019 public water supply systems in the State. Approximately 60 per cent of these were municipally owned, about 40 per cent privately owned, as noted above. They served 1,501 com-

---

to 45 gallons per minute, or about 65,000 gallons a day. (Chap. 135, Laws of 1954.)

[14] Conservation Commission, *Pure Water: Plenty of It in Every Home and Factory* (Albany, n.d. but about 1912).

[15] Conservation Commission, *Seventh Annual Report* (1917), p. 85.

[16] In 1952 the State Comptroller appointed a committee to study the legal, financial, and practical barriers to the development of public water supplies and to the distribution of water to those who need it. The report of the committee treats of these and many like problems. See *Report of State Comptroller Committee on Problems of Water Distribution* (1953).

munities with a total population of 14,130,192.[17] Total consumption for
the State from all public supplies was estimated at 2.14 billion gallons
per day.

Water Quality

The problems of quality and quantity have gone hand in hand as
companion concerns of the State with respect to water resources;
it was the breakdown of quality, indeed, which led the State to its early
expressions of interest in New York City's water problem. That interest
remained unabated throughout the last century. Thus in 1866 the
Legislature passed an act creating a Metropolitan Board of Health
with power to order the correction of any unsanitary condition or practice which it regarded as ". . . dangerous or detrimental to life or
health, . . ." (Chap. 74, Laws of 1866).

Statutes of state-wide application in the public health field go back
to well before 1800; but insofar as the early acts dealt with sanitation,
they merely endowed the cities, villages, and towns with authority to
enact sanitary regulations. It was not until 1880, when the first Public
Health Law (as such) was passed (Chap. 322, Laws of 1880), that the
problem of sanitation was approached on a broad front. Even then,
the attack was hesitant and exploratory, as is evidenced by the first
public utterance of the newly created State Board of Health:

> The want of adequate laws for preventing the pollution of streams,
> wells, and other sources of supply of water is equalled only by the
> general neglect of the duty to avoid the domestic use of sanitarily im-
> pure and defiled water. The duty of providing pure supplies and of
> maintaining necessary local regulations to prevent the sources of defile-
> ment is now under discussion in various parts of the State, and is
> receiving the special attention of this Board. . . . In furtherance of
> the design to make good use of all existing laws for sanitary protection,
> the Board has issued to all local authorities a proposed form of sanitary
> regulations upon this subject.[18]

It will be noted that the Board continued to place primary faith in
local authorities, as had been the custom throughout the century. The
normal procedure for the correction of unsanitary conditions called
for the declaration of the existence of a nuisance by the local authority

---

[17] New York State Department of Health, Bulletin 19, *Public Water Supply Data*
(1954), pp. 4-5.
[18] Board of Health, *First Annual Report* (1881), p. 21.

(usually a board of health), followed by the usual legal steps for the abatement of a nuisance.

Between 1880 and 1892, the Legislature passed a number of laws designed to protect public water supplies, and in 1893 that body codified the public health laws. A number of specific acts passed during the years immediately following proved unequal to the task of dealing with the mounting problem of pollution, which by 1900 or a little after came to be recognized as one requiring special attention. By that time, most of the State's streams had become polluted, many of them seriously, by municipal sewage and industrial wastes. These conditions led to the enactment in 1903 of the first general anti-pollution law (Chap. 468, Laws of 1903), which with remarkably few amendments remained the basic public health law on the pollution of waters for almost half a century. The act prohibited the discharge into public waters of sewage and waste in quantities sufficient to injure public health without the approval of the State Health Commissioner.

Meanwhile, another kind of interest in water pollution was taking form with the emerging concern for natural resources. This concern found expression first in a legislative act relating to fisheries, then in a law creating the Forest Preserve. The latter, through recognition of the importance of watershed protection, in a sense dealt with water resources. Shortly after 1900 several statutes dealing with water as such were passed. Between 1900 and 1913 the Legislature passed a number of individual acts which in the latter year were consolidated into a pollution law. Since the act concerned conservation, it emphasized fish life, prohibiting the discharge into any waters (private or public) of wastes which would be injurious either to fish or to conditions necessary to fish propagation. Like the earlier health law, this statute remained substantially unchanged for many years.

Two basic statutes, then, on the subject of water pollution came into being. One concerned pollution which affected health, and was administered by the Department of Health; the other, relating to pollution which affected fish life, was administered by the successive conservation agencies.

Pollution control for almost the whole of the half-century to 1950 was haphazard, uncoordinated, and generally ineffectual, and progress toward abatement was slow. Both health and conservation agencies conducted investigations of stream pollution, and by 1940 some 45 studies of the subject had been published. Further, both fought within their means to ameliorate stream pollution, but the growth of the cities and the development of industry made it difficult for them to stay abreast of the problem. The hand of the Department of Health was strengthened with the creation, in 1913, of a Public Health Council with authority to promulgate a Sanitary Code (Chap. 559, Laws of 1913). The Conservation Commission experimented with new forms of organization for pollution control; and in 1918 the Conservation Commissioner invited Dr. Henry B. Ward, Professor of Zoology in the University of Illinois, to make a study of the pollution problem and to assist in outlining a permanent policy for the progressive cleaning up of the State's waters.[19] Five years later the Conservation Commission, in a special report on stream pollution, voiced something approaching despair. Analyzing the statute bearing on stream pollution, the report discovered ". . . dispersed and incomplete authority, lack of precision and clear-cut definition, lack of uniformity and power to enforce regulations, and lack of coordination." Further, the report found three state departments with jurisdiction over stream pollution problems. "Necessary regulation can best be brought about," the report concluded, "by the creation of a special Commission clothed with wide quasi-legislative powers enabling it to investigate conditions, allocate different waters for different purposes and make regulations to bring this about." [20]

Meanwhile, the Legislature was not idle. It was declared state policy to control and abate stream pollution in the interest both of public health and of fish life; and the Legislature passed frequent acts looking toward the implementation of that policy. Further, it made an occasional modest appropriation for a special study. A tangible evidence of legislative interest is found in a joint resolution calling for the appointment of a Joint Legislative Committee in Reference to the Pollution of Waters in the State. Passed April 10, 1924, the resolution charged the committee to look into the problem of pollution in and around New York harbor, and also, significantly, to ". . . examine the sewage systems and methods of sewage disposal in the cities of this State for the purpose of modernizing the same, of suggesting plans for such purpose and recommending such method of sewage disposal as may be proper to protect human life and health." A second joint resolution extending the life of the Committee gave it the power to investigate also the pollution of waters by industrial wastes.[21] Notwithstanding the intent of the Legislature, the Committee confined its study to the New

---

[19] Conservation Commission, *Eighth Annual Report* (1918), p. 154. See also Henry B. Ward, *Stream Pollution in New York State* (Albany, 1919).

[20] Conservation Commission, *Thirteenth Annual Report* (1923), pp. 236-239.

[21] The preliminary report of the Committee, published as Legislative Document, 1926, No. 86, contains these joint resolutions, at pp. 3-5.

York City area, whose waters it found to be fearfully polluted. As a postscript gesture in recognition of its wider responsibility, the Committee reported badly polluted waters from both municipal and industrial wastes upstate, particularly in the Mohawk and Hudson Rivers.[22]

The most significant step in behalf of stream pollution taken to that date came about through the creation, in 1935, of an agency whose principal concern, nominally at least, was not water pollution at all. The agency in question, the Joint Legislative Committee on Interstate Cooperation, nevertheless found itself from the beginning involved in the problem of pollution of interstate waters and so, by easy transfer, of intrastate waters as well. The Committee may well count its principal achievements to be state adherence to several arrangements for the abatement and control of pollution in interstate waters; but it played an important role nevertheless in bringing about effective pollution control in domestic waters as well.

The Legislature recognized the Joint Legislative Committee on Interstate Cooperation as a potential source of vigorous leadership in pollution control, and on March 26, 1946, adopted a concurrent resolution directing the Committee ". . . to make a study of the subject of pollution as to all of the waters within the state and to conduct such study in such manner as it, in its wisdom and experience, believes to be most efficient, effective and intelligent in the interest of the people's welfare and to report its findings on this subject to the Legislature on or before the 15th day of February, 1947, in a separate report." [23] In pursuance of this resolution, the Chairman of the Joint Legislative Committee appointed a Special Committee on Pollution Abatement. The Special Committee, though established on a one-year basis (as, indeed, was the parent body), continued in existence for five years, during which it subjected its central issue to vigorous and continuing scrutiny. The five reports which it published during that time focused attention sharply on the problem of pollution.

In 1951 the Legislature separated the problems relating to natural resources from those more broadly interstate in character and created the Joint Legislative Committee on Natural Resources to assume jurisdiction over the former. The concurrent resolution creating the new

[22] *Report of the Joint Legislative Committee in Reference to the Pollution of Waters of the State* (Legislative Document, 1927, No. 78), especially pp. 32-33.

[23] *Progress Report of the Special Committee on Pollution Abatement of the Joint Legislative Committee on Interstate Cooperation* (Legislative Document, 1947, No. 59), p. 13.

Committee (adopted March 14, 1951) specifically mentioned the water pollution problem and charged the Committee with responsibility therefor. The Joint Legislative Committee on Natural Resources has carried on in the tradition of its predecessor. It has not centered attention on the water pollution problem quite so sharply as did the preceding Special Committee on Pollution Abatement, but it has included in each of its five reports a substantial section on the subject. For eleven years (1946-1957), then, the State has had a joint legislative committee with a primary concern for water pollution, and the subject has been quite thoroughly aired during that period.

The Special Committee on Pollution Abatement faced a formidable task at its launching in 1946. A 1940 survey of the U.S. Public Health Service had identified New York as facing the greatest sewage and waste treatment problem of any state; Pennsylvania, ranked second, had a pollution task only half as great as New York's.[24] Weaknesses in the State's pollution program early noted by the Committee as requiring legislative action included divided authority over pollution control, gaps in the law, after-the-fact rather than preventive action, inadequate control over industrial wastes, and the absence of a unified system of stream classification.[25] The Committee immediately recommended, as a first step, that the Legislature offer state aid to public units in the planning of sewer systems and sewage treatment works by granting matching funds for planning up to the limit of 2 per cent of the estimated construction cost; and the legislative body acted favorably on this recommendation in 1947 (Chap. 578, Laws of 1947).

The big job, however, was that of drafting an inclusive anti-pollution bill, a task to which the Committee addressed itself early in its second year. It proceeded by requesting the Commissioner of Health and the Commissioner of Conservation to develop a joint statement of the desirable scope and content of such a measure; then it invited those officials to prepare a draft, with the assistance of a legal expert, in accordance with the statement of basic concepts as approved by the Committee. Subsequently, the draft, approved by the Committee, was presented for public discussion at a number of conferences held over the State, at which municipal and industrial representatives were asked to present their reaction to the proposed measure.[26] The result of this careful groundwork appeared when, in 1949, the Legislature passed

[24] *Ibid.*, p. 27.

[25] *Ibid.* (Legislative Document, 1948, No. 50), p. 55.

[26] The procedure followed is described in *ibid.*, pp. 56-57, and *ibid.* (Legislative Document, 1949, No. 51), pp. 29-31.

the Water Pollution Control Law, as recommended, without a dissenting vote in either house (Chap. 666, Laws of 1949).

The Act began with a declaration of policy (Sec. 1200 of the Public Health Law) which read as follows:

> It is declared to be the public policy of the state of New York to maintain reasonable standards of purity of the waters of the state consistent with public health and public enjoyment thereof, the propagation and protection of fish and wild life, . . . and the industrial development of the state, and to that end require the use of all known available and reasonable methods to prevent and control the pollution of the waters of the state of New York.

Another significant policy statement declared that

> It is the purpose of this article to provide additional and cumulative remedies to abate the pollution of the waters of the state and nothing herein contained shall abridge or alter rights of action or remedies now or hereinafter existing, nor shall any provision in this article . . . be construed as estopping the state, persons, or municipalities . . . in the exercise of their rights to suppress nuisances or to abate any pollution now or hereafter existing (Sec. 1260).

The Law provided for a Water Pollution Control Board, whose organization and method of operation are described below in Chapters III and VII respectively.

Aside from its part in bringing about the Water Pollution Control Law, which it rightly regarded as its principal contribution, the Special Committee and its successor (the Joint Legislative Committee on Natural Resources) have pursued several significant activities. First, the Special Committee encouraged state participation in a number of interstate arrangements for pollution abatement. Second, both Committees have interested themselves in certain substantive pollution problems, as for example the abatement of pollution arising from the Long Island duck farms and of industrial pollution of the Buffalo River. Third, the Committee (regarding the two appearing in series as one) has considered itself as both the public and legislative liaison agency for the Water Pollution Control Board, and has conducted itself accordingly; for example, it has cooperated with the Board to work out amendments to the pollution control law and has shepherded the amendments through the Legislature, and it has called public conferences at the instance (or with the approval) of the Board. The close relation between the Committee and the Board is attested by the fact that the annual report of the latter appears as a part of the Committee's yearly report. Fourth, the Committee has kept a close eye on Federal interest in pollution control, and has brought its influence to bear in support of Federal legislation on the subject. The Committee noted with satisfaction that Public Law 845 (80th Congress, 1948) represented the first measure for Federal participation in pollution control after congressional rejection of 107 bills over half a century.[27]

Finally, the Committee has recognized that, notwithstanding its anticipated beneficent effects in the end, the Water Pollution Control Law will result in immediate hardships for large numbers of cities and industries. Literally hundreds of cities and industries do not now meet reasonable standards in the disposal of their wastes, and the Committee perceived early that the real bottleneck in the enforcement of standards would inhere in the financial outlays involved. The Committee foresaw that the new Water Pollution Control Board would receive no more in the way of cooperation than its subjects would be able to pay for; and the bill would be high: the Committee estimated as early as 1949 that the new Act in the end would cost the municipalities more than $400,000,000, and industries a like sum.[28]

The Committee undertook to cushion the impact of the law by seeking to help the municipalities and industries find ways of meeting their new obligations. Harkening to the plaints of local officials and their organizations, it concerned itself particularly with the problems of cities, towns, and villages. It cooperated with the State Comptroller's Committee on Problems of Water Distribution; and it conducted a number of studies and made several recommendations on its own account regarding both water supply and sewage disposal. A number of steps to liberalize the law, both constitutional and statutory, resulted from these efforts. Among these was a constitutional amendment, approved by the people in November, 1955, permitting municipalities to construct joint sewage facilities (as well as drainage facilities) to serve two or more units of government, and allowing them to construct such facilities and then to contract with neighboring communities to take care of their sewage and drainage needs. Another measure, adopted in 1954, authorized the establishment of county sewer districts; and yet another extended to such districts the state aid already enjoyed by municipalities in the preparation of plans for sewage facilities.[29]

---

[27] *Ibid., Report* (1949), pp. 41-43.

[28] *Ibid., Report* (1950), p. 19.

[29] These measures and others like them (both successful and unsuccessful) are described in *Report of the Joint Legislative Committee on Natural Resources*

These acts recognized and sought to deal with a serious problem. That they were not successful is indicated by a report which appeared in the *New York Times* on July 28, 1957, on the subject of the fiscal capacity of the municipalities to comply with the orders of the Water Pollution Control Board. The article concerned a challenge of a Board classification by the Village of Waterford. The Supreme Court in this instance upheld the Board. The article goes on to state that compliance with Board orders would cost Utica $5,600,000, Amsterdam $3,000,000, and Newburgh $2,800,000, as three illustrations of many distressed municipalities. On July 29, 1957, the *Times* reported a proposal by Assembly Speaker Oswald D. Heck to establish a state fund of $500,-000,000 (by bond issue) from which low-cost or interest-free loans would be made to communities required to improve their sewage treatment plants or to install new facilities. Meanwhile the Water Pollution Control Board proceeds apace with its task of classifying the State's waters under the Act of 1949. The legal tide is running with the Board, and so is public opinion; but the problem of reconciling large new capital expenditures with limited financial capacity, and particularly with municipal debt limits, is one that has not thus far been solved.

Water Power

New York's hydroelectric power potential has been recognized since the advent of electricity.[30] This resource is estimated at 5,390,000 kilowatts, or somewhat less than 5 per cent of the Nation's total; of this, after a full half-century of travail only 19 per cent was developed as late as 1956. The ratio of developed to undeveloped power will of course be drastically changed by the construction now under way on the St. Lawrence and that in prospect for the Niagara.

The State has always had a direct interest in this important natural resource, and for two reasons. First, when the value of hydroelectric power sites first began to be recognized, there was no fixed state policy regarding private as against public development; and there were many and powerful voices who maintained that *all* water power should be developed by the State. Second, New York owned or had proprietary

---

(Legislative Document, 1955, No. 76), pp. 143-149; and *ibid.*, (Legislative Document, 1956, No. 63), pp. 237-238.

[30] A. Blair Knapp, *Water Power in New York State* (Syracuse: Syracuse University, 1930), pp. 48-50. This study constitutes an excellent summary of water power developments in New York to 1930.

rights over many important power sites. This was true, for example, with regard to the sites located within the Forest Preserve, the power latent in surplus canal waters, and the immense hydroelectric potential in the international rivers, concerning whose development the State had an important if not a controlling voice. One way or another, therefore, New York was almost of necessity in the power business. How it should discharge its responsibilities was a major subject of public debate for almost half a century.

We have observed that one of the early laws of general application dealing with water resources was that of 1905 which established the Water Supply Commission. That body turned its attention at once to the subject of water power, as it did indeed to various other aspects of the water problem.[31] Its jurisdiction was placed on a much firmer footing, however, by two significant events of 1907. First, in his annual message to the Legislature in January of that year, Governor Charles E. Hughes expressed the view that water power ". . . should not be surrendered to private interests but should be preserved and held for the benefit of the people," and called for action to define a state policy on the subject. In response to his recommendation, the Legislature passed the Fuller Act (Chap. 569, Laws of 1907), by which the State adopted a forthright stand. Section 1 of the Act proclaimed the State's policy in these words: "After the passage of this act, the state water supply commission is authorized and directed to devise plans for the progressive development of the water powers of the state under state ownership, control and maintenance for the public use and benefit and for the increase of the public revenue." The Commission was admonished to proceed without delay, to the end that a plan might be presented to the Governor and the Legislature at the earliest possible moment. The Fuller Act took effect even as the nation, under the leadership of President Theodore Roosevelt, was becoming conscious of natural resources and the need for conservation.

The Water Supply Commission employed a consulting engineer whose work enabled the Commission to present a substantial, if admittedly incomplete, progress report on February 1, 1908. The report included the following significant passages:

> Single interests which have studied the subject and become familiar with the value of water rights and the economy of water over steam, and who look only to private gain, may and probably will raise objections to any action on the part of the State toward ownership of water

[31] State Water Supply Commission, *Second Annual Report* (Albany, 1907), pp. 16-17.

power rights. It is clear, however, that individuals cannot economically develop the water power within this State and the opportunity for the State to do so should not be defeated by the determination of a few intent upon amassing fortunes and creating monopolies. Such purposes must yield to that broad policy which will most evenly and justly distribute the wealth of the State.

To acquire the dam sites and storage rights on the large streams and begin the systematic building of storage dams at such places as will most quickly earn a revenue that will repay to the State the cost of acquiring such rights and of the necessary construction seems to the Commission to be the plain duty of the State.[32]

This statement embodied the policy toward which the Commission and its successor agency labored for the next several years.

The Water Supply Commission pursued the directive fixed by the Fuller Act to the end of its life. In its last official report, the Commission gave voice to the conviction which had guided it in these terms: "The trend of public sentiment in favor of State ownership, development and control of water power resources has been unmistakably indicated during the last few years. The agitation for the establishment of this principle has been widespread and has met with a ready response." [33] Notwithstanding the Commission's optimism, a bill introduced in 1910 to effectuate its program failed to command any substantial support in the Legislature.

The Water Supply Commission was superseded in 1911 by the Conservation Commission, which under the new conservation law took up where the predecessor agency had left off. It continued with both the water power surveys and the support of state development of its water power resources which had characterized the earlier agency. In 1912 a measure called the Bayne Bill, designed to carry into effect the major recommendations of the Commission, received a favorable vote in the Senate but failed to pass the Assembly. Other bills during the next two years met the same fate, notwithstanding continued vigorous support of a program of public action by the Conservation Commission.

Clearly the policy of public development outlined by the Fuller Act in 1907 had failed to find the support necessary to its effectuation in positive action. The reason for this was simple enough: the question of public versus private development of the State's water power resources had emerged over the years as a political issue, with the Demo-

[32] State Water Supply Commission, *Progress Report on Water Power Development* (Albany, 1908), pp. 27-28.
[33] State Water Supply Commission, *Sixth Annual Report* (Albany, 1911), p. 56.

crats identified as the proponents of public power and the Republicans as the champions of private development. By 1915 the lines were clearly drawn, and there was no turning back. When the Republicans gained control of the Conservation Commission in that year, they were able at last to reverse the verbal policy, as they had been able all along to defeat translation of that policy into effective action.

The Machold Storage Law which was passed by the Legislature in 1915 (Chap. 662, Laws of 1915), clearly reflected the capture of state power policy by the Republicans. The act provided for the creation of river regulating districts, which were empowered to construct storage reservoirs for the control of stream flow. Significantly, there was no mention of the state ownership and development of power sites that had been so vigorously supported for eight years; instead, the law emphasized a new philosophy of private development. So did the annual reports of the Conservation Commission for the several years following 1915. That of 1916, for example, stated that "The Commission does not believe that the time has arrived when the State should undertake the vast project of supplying its people with power." It follows, therefore, the Commission went on, that the State should place its reliance in private development, "renting" the power ". . . to private individuals or companies on such terms as will yield a return to the investor adequate to the conditions presented under each separate development, . . ." [34] The succeeding report contained this thought:

> For some time past the inventory and preliminary studies of the water powers of the State have reached a point where the vast resources of the State in this most valuable natural asset have stood fully revealed. In the face of this fact and the overwhelming need for power at the present time, we can no longer, without serious danger of becoming economic slackers, sit with folded hands while the power of our streams runs uselessly to waste.[35]

Clearly the Commission had had enough of studies; it wanted action. The next several annual reports emphasized the intention of the Commission to favor private over public development, removing any vestiges of doubt as to the way the tide was running.

With the consolidation of their positions in Legislature and administrative agencies, the advocates of power development by private enterprise began to agitate for an official new statement of public

[34] Conservation Commission, *Sixth Annual Report, 1916* (Albany, 1918), p. 40.
[35] Conservation Commission, *Seventh Annual Report, 1917* (Albany, 1919), p. 80.

policy in the power field. An opportunity for positive action was provided by the example of Congress in 1920 in passing the Federal Power Act, which was interpreted in New York as an invitation to like action by the State. Alfred E. Smith, who was Governor in 1919 and 1920, vigorously resisted the movement for a new law; but he was overwhelmed when, in 1921, the Republicans elected majorities in both houses of the Legislature and a Governor as well. The trend could have but one outcome: in 1921, the Legislature passed and the Governor signed a new Water Power Act (Chap. 579, Laws of 1921). Republican leaders hailed the new law as embodying a positive state power policy, adopted after twenty years of valiant effort. In truth it had precisely the same status as the Fuller Act of 1907 which it superseded: it constituted a verbal expression of legislative intent, nothing more. It remained to be seen whether or not the expression would be implemented as policy.

The essence of the new law was found in its provision for licensing the State's water power sites to private individuals or corporations, who would pay annual rentals for their rights and who would, of course, be subject to public control. The Act also established a Water Power Commission, which proceeded about its task without delay by granting preliminary permits for two St. Lawrence River projects, ". . . the first favorable action on the various applications before it." The new commission noted with satisfaction that the proposed projects would generate upward of 1,000,000 horsepower of electricity for distribution in New York State.[36]

Passage of the Water Power Act ensured an intensification of the struggle between the advocates of public power development and those of private ownership. The new phase of the controversy was to last for a full ten years. The Republicans, in control of the Legislature and the administrative agency principally concerned for most of the decade, worked unceasingly for private development. The Democrats, led by Alfred E. Smith, who as Governor for most of the period occupied a strategic position, threw up road blocks and resisted the Republican maneuvers at every turn. Public power scored a minor victory when in 1922 the Legislature authorized state construction of power plants at two canal sites, Crescent Dam and Vischer Ferry Dam (Chap. 532, Laws of 1922); but that proved to be a limited victory when, on completion, no municipality appeared prepared to take the power generated, which thereupon was contracted to private companies. Except

[36] New York Water Power Commission, *First Annual Report* (Albany, 1922), pp. 43-44.

for this single light, the way of public power was dark indeed during the early 'twenties. The cause may have been saved from oblivion by the executive reorganization of 1927, which provided for a Water Power and Control Commission to be composed of the Conservation Commissioner, the Superintendent of Public Works, and the Attorney General. It will be observed at once that the Governor would be in a position to control the new Commission through appointment of two of its members. The new law also provided that no license issued by the Water Power and Control Commission ". . . shall be effective unless and *until it is approved in writing* by the governor and such approval is signed by him and affixed thereto; . . ." (*Conservation Law*, Art. 14, Sec. 615).

With the approach of the effective date of the administrative reorganization, the Water Power Commission rushed through the final stages of granting licenses for the private development of St. Lawrence power. Governor Smith warned the Commission of its impending doom, and in a series of communications emphasized what he termed the doubtful legality of the step contemplated. The resulting furor was sufficient to frighten away the prospective licensees, and the move for the development of St. Lawrence power by private enterprise fell to the ground.

Meanwhile, Governor Smith as early as 1924 had proposed a plan for a state power authority to be charged with development of the State's water power resources. In brief, the authority as conceived by the Governor would construct, own, and operate the facilities necessary to bring to full development the water resources of the State. The Governor urged his plan upon the Legislature repeatedly; and more than one bill was introduced to create an authority, only to be uniformly rejected by the Republican-controlled Legislature.

In the gubernatorial campaign of 1928 the Democratic candidate, Franklin D. Roosevelt, energetically espoused the Smith public power program. His Republican opponent as vigorously championed the cause of private development, and the issue was squarely joined before the electorate. The election of Roosevelt presaged action on the issue, action which took the form of an act passed by the Legislature in 1930, establishing the St. Lawrence Power Development Commission (Chap. 207, Laws of 1930). The new law provided (in Section 1) that

> The natural water-power sites in, upon or adjacent to the Saint Lawrence River owned or controlled by the people of the state or which may hereafter be recovered by them or come within their owner-

ship and control, shall remain inalienable to, and ownership and control shall always remain vested in the people of the state.

On January 15, 1931, the Commission submitted to the Legislature a report presenting economic, engineering, and legal justifications for the development of the St. Lawrence power potential by the State. That body responded by passing, without delay and without a dissenting vote, an act providing for the Power Authority of the State of New York (Chap. 772, Laws of 1931). Once again the Legislature had declared its intention to establish a state policy regarding water power resources; and once again it remained to be demonstrated whether or not that policy would find expression in effective action. The story of the Power Authority and its twenty-five-year struggle to bring to fruition the policy declared in 1931 is related in Chapter VIII.

Water Control

In a sense, each of the four subjects discussed to this point—navigation, water supply, water quality, and water power—is related to the larger problem of water control; for control of water lies at the center of most water programs. It is proposed to limit the present discussion to water control somewhat more narrowly defined to include river improvement, river regulation, and flood control. The definition by design excludes the management of land for water control purposes.

These subjects have received varying treatment at the hands of the State for upward of a century, with one term in use at one time and another at another. Whatever the designation or the avowed aim of a particular water control law or program, however, a number of secondary purposes usually crept in. It remained for the Joint Legislative Committee on River Regulation as late as 1952 to point out the essentially multiple-purpose character of a water control program. The principal ends to be served by river regulation, the Committee observed, are flood prevention, water supply for municipalities and industries, power production, low-flow regulation, water storage (for fire-fighting, irrigation, and domestic use), recreation, storage of reserve water for canals, and service as water basins to restore ground water sources.[37] Here was a broad conception of the multiple functions to be performed by a water control program. Not all control programs

[37] Joint Legislative Committee on River Regulation, *Report* (Legislative Document, 1952, No. 51), pp. 18-21.

served all of the purposes named; but some served several, and almost all served more than one.

Early interest in impounded water in New York centered on the need for water sources by the cities and for "feeder" waters for the canals. As early as 1854 the Legislature appropriated $31,250 for river improvements, which caused the State Comptroller to suspect darkly that the action was designed to serve private gain under the cover of public interest.[38] The first storage reservoir for flow regulation was Cranberry Lake on the Oswegatchie River, constructed in 1867. Between that date and the end of the century, several additional reservoirs were constructed, all of them for stream regulation or for compensatory water flow and all under individual and separate statutes.

Following serious floods in 1902, a Water Storage Commission was appointed to look into the subject of water storage with an eye to flood prevention. The Commission, citing what it called the "injurious irregularity" of stream flow, recommended that river regulation be placed in the hands of a permanent commission, with power to plan and build the necessary structures and to assess the costs against the beneficiaries. As a result of this recommendation the Legislature in 1904 passed the River Improvement Act (Chap. 734, Laws of 1904), which created a River Improvement Commission vested with powers to regulate the flow of rivers and water courses. The Commission was composed primarily of high state officials, who soon found their *ex officio* duties heavier than they were able to discharge satisfactorily. Accordingly in 1906 the River Improvement Commission was abolished and its responsibilities transferred to the newly established (1905) Water Supply Commission. The latter agency pursued its functions with regard to river improvement with considerable vigor, as did the Conservation Commission which took over in 1911.

For all their efforts, however, little was accomplished under the River Improvement Act. Several petitions (locally drafted, circulated, and signed, as required by law) were filed for improvement projects, but only three reached the stage of Commission action. Two of these failed of approval (one failed to be approved by the Commission, one, recommended by the Commission, failed of approval by the Legislature), so that only one application succeeded in clearing all official hurdles. That one called for improvement work (channel straightening, dredging, ditching) on Canaseraga Creek. Approved by both Commission and Legislature, the proposal faced vigorous challenge in the courts before final sanction to proceed was gained. Then there were difficulties with

[38] State Comptroller, *Report, 1855*, p. 61.

financing. Finally all obstacles were removed, and in 1911 work began on the Canaseraga project, which was completed in 1916. According to official reports, the improvement was highly successful. Nevertheless the Water Control Commission reported in 1922 that, because of lack of funds, it had not been able to do any maintenance work on the Canaseraga Creek project during the preceding year.[39] Thus ended, through simple non-use of the instrument provided, a not overly bright chapter in New York's history of river improvement.

But what had failed to find takers as a river improvement program might succeed under another name. Note has been made (in the section on water power, above) of the shift in state policy which occurred in 1915, and of the Machold Storage Law which gave legislative endorsement to the new policy. The Act provided for the creation of local (river drainage basin) river regulating districts, whose stated purposes were the control of floods and the increase of low stream flows through construction and operation of storage reservoirs.

Under the Machold Act several petitions were presented to the Conservation Commission and its successor in this field, the Water Control Commission, seeking the establishment of river regulating districts. Of these, all but two were withdrawn, usually following protests by local interest groups. Two applications were allowed, the first for the establishment of the Black River Regulating District (1919), the second for the creation of the Hudson River Regulating District (1922). The Black River district, after successfully withstanding an attack in court, proceeded to enlarge the Stillwater reservoir, on which work was completed in 1925. The Sacandaga reservoir had long been a favorite project of those interests in the upper Hudson, including the City of Albany; the Hudson River district, likewise successful in withstanding a judicial assault, undertook construction of the Sacandaga reservoir, which was finished in 1930. The vicissitudes of the Black River Regulating District in its efforts to bring one of the State's major streams under control are recounted below in Chapter VI.

[39] Water Control Commission, *Report, 1922* (Legislative Document, 1923, No. 93), p. 3. The failure of the Canaseraga Creek project is attributed by a prominent state official to the sequence of the steps taken. Construction was completed *before* assessments were made; the assessments then were reviewed by the court, which handed down an adverse ruling. The decision was not appealed, the beneficiaries refused to pay their assessments, and the State had to meet all the costs of the improvements. Since the Canaseraga experience, it has been virtually impossible to sell "river improvement" bonds. The situation might (the official in question believes would) have been entirely different had the problem of assessments been worked out before the construction was undertaken.

Here again a program for river control which initially appeared to hold promise in the end failed to produce widespread results. The history of storage reservoir construction by the State is, indeed, far from impressive. The approximately ten reservoirs constructed by the State (or under State auspices) since 1867 constitute no more than a minor fraction of the structures necessary to rid New York of the losses consequent on unregulated stream flow.

Fortunately the story does not end with the ineffectual efforts thus far recounted. A devastating flood hit south and central New York in July of 1935, leaving in its wake forty-three deaths and damages estimated at $40,000,000. This catastrophe helped Congress arrive at a favorable decision on the Federal Flood Control Act, which became law in 1936. It also prompted the New York Legislature almost simultaneously to pass a State Flood Control Act (Chap. 862, Laws of 1936). The main import of the Federal Act was the assumption by the national government of responsibility for flood control on navigable rivers and their tributaries. The Act placed the main burden of Federal responsibility on the Corps of Engineers of the U.S. Army, though it invited active cooperation by the states, which were, indeed, required to pay for the properties required and the damages resulting from Federal flood control projects located within their boundaries. The New York law created a State Flood Control Commission of twelve members to serve as the State's liaison agency with respect to the Federal program. The Commission has no operating responsibility, but represents the State in all negotiations relating to the Federal flood control program. The state Department of Public Works is charged with the duties of acquiring the necessary properties for flood control projects in the first instance and of managing completed projects turned over to the State in the second. Its managerial obligations have not proved onerous thus far, since only one structure (the Syracuse dam) has been turned over to the State for operation and maintenance.

The twenty-two-year-old flood control marriage between the State of New York and the Federal government has been a happy one, as related by the State Flood Control Commission. The annual reports of the Commission reflect a zestful approach to its duties, a healthy respect for the Corps of Engineers, and a sense of satisfaction with the results achieved. The results have in truth been quite substantial. Since 1936, 73 flood control projects have been authorized, of these (to 1956) 56 have been completed. The projects authorized include 11 dams and 62 local improvements. The 11 dams are estimated to cost more than $96,000,000 when completed, of which the State will be required to contribute less than half a million dollars. The local improvements are estimated at somewhat more than $50,000,000, of which the State and localities are expected to contribute about $6,500,000. Of the total of almost $149,000,000 (there are a few costs not represented in the construction cost figures above), the State's share will be less than $7,700,000. The Commission speaks of the gross figure as representing ". . . a Federal-State capital investment . . . ," as indeed it does; but it has a sort of mule:rabbit appearance.[40] In terms of results achieved and comparative contributions, it is fortunate for New York that this program evolved without worry about the relative responsibilities of Federal and state governments for flood control. It would have cost the State a great deal of money to have remained loyal to lofty principle here.

A newcomer to flood control work in New York is the United States Department of Agriculture, which operates in this domain principally through the Soil Conservation Service. The SCS in the past has concerned itself chiefly with land treatment measures and small (upstream) water retardation structures. It is likely that its activities will increase substantially under the impetus of the currently expanding small watershed program, about which more will be said in another connection.

Here, then, is a rousing success story, one which began inauspiciously with ineffective programs of river improvement and river regulation but which found its feet in a cooperative program for flood control. That the program achieved success only with the entry of the Federal government, and that, money-wise, it is 95 per cent Federal and only 5 per cent State are sobering facts which are worthy of further contemplation.

Miscellaneous

Among the several miscellaneous water programs worthy of brief mention are those which involve the drainage of swamp, bog, and other low or wet lands. The State's acreage in such lands is variously estimated at from 270,000 to 300,000, much of it susceptible of improvement which would make of it fertile agricultural land. There have been drainage laws since the early days of statehood; and a great deal of reclamation work was done under those laws before 1900, but

[40] Flood Control Commission, *Twenty-First Report* (Legislative Document, 1956, No. 54), p. 40. For a list of all projects authorized and approved, see *ibid.*, pp. 43-44.

without notable success. Drainage work in those days was placed in the hands of commissioners appointed by the county courts. They were local men, chosen usually without regard for engineering or other technical qualifications—oftentimes, indeed, they were frankly political appointees. The principal results of their labors are to be found in abandoned "commissioners' ditches" traversing the major swamps, monuments to an ineffective system.[41]

It remained for the Conservation Law of 1911 (Chap. 647, Laws of 1911) to begin to bring some sort of order out of the drainage chaos. The new law rested upon the concept of state action under the police power wherever drainage should prove necessary for ". . . either the public health or the public safety, or the public welfare, or all or any of them." The newly created Conservation Commission was made the agent of state action, and was authorized to proceed with drainage work either on its own initiative or at the request of interested local units or parties—cities, counties, towns, villages, or individual or associations of individual property owners. Whatever the inspiration for action, the Commission was directed to conduct investigations and hearings to determine both the need for drainage work and the attitude of the affected property owners regarding the proposed improvement. Ultimately, if the Commission found in favor of action and the Governor approved, a drainage improvement district would be established. Work then would proceed under the direction of the Commission; it would be financed by the sale of serial bonds secured by taxes laid upon the property of the district. The drainage law has been amended a number of times and elaborated in some respects, but it remains in general outline substantially as it was passed in 1911.

Under the drainage law the Conservation Commission, and after it the Water Control Commission (1922-1927) and the Water Power and Control Commission (1927 to date), has received a great many petitions for the establishment of drainage districts. Out of a welter of discussions and hearings covering some 45 years, five drainage improvement districts emerged between 1930 and 1942. The total acreage covered by the five is somewhat less than 29,000, and almost two-thirds of this lies in a single district (No. 3) which is by no means fully developed. The outstanding (bonded) debt of all districts is a little more than $38,000, although three have no debt at all. The districts are administered by the Water Power and Control Commission direct.[42]

[41] Conservation Commission, *First Annual Report,* II (Albany, 1912), p. 43.
[42] Conservation Department, Division of Water Power and Control, *Twenty-Eighth Annual Report* (1954), pp. 266-270.

An amendment to the Conservation Law provides a method by which drainage improvements may be effected by the cooperative action of a group of landowners whose property is in need of drainage. Such groups may establish themselves as drainage section associations, in which guise they enjoy certain rights of action by state law. State administrative supervision is nominal, consisting of nothing more than the requirement that certified copies of the articles of association be filed with the Conservation Department. Another provision of the law permits either drainage section associations or individuals to acquire easements in the property of others necessary to effectuate drainage projects on their own property. Still another provision authorizes acceptance of Federal aid in the creation and operation of drainage districts, and details the procedure to be followed to that end.[43] High hope was held for this projected program, but to date nothing has come of it.

A second miscellaneous program concerns the Saratoga mineral springs. These springs, originally privately owned and exploited, were all but exhausted by the beginning of this century, when public demand arose for their preservation. The Legislature responded by passing an anti-pumping law in 1908; and the next year it took the additional important step of designating Saratoga Springs a state reservation and providing for the public acquisition of the properties there. During the next half-dozen years title to most of the springs passed to the State, and in 1916 the Legislature passed a law placing the Saratoga Springs Reservation under the jurisdiction of the Conservation Commission (Chap. 296, Laws of 1916). The Commission hailed this step as ". . . one of the most important acts of conservation ever undertaken by the State of New York." [44]

The State has been in the business of operating a mineral springs spa, then, for almost half a century. To make explicit the policy implied in state acquisition of the Saratoga properties, the Legislature in 1930 declared the development of the State Reservation there to be a part of the State's public health policy (Chap. 866, Laws of 1930). The management of the Reservation nevertheless was left in the hands of the Conservation Department, which has assigned this responsibility to a special division. Among the enterprises managed by the division are bath houses, a drink hall, a swimming pool, tennis courts, a golf course, parks and picnic grounds, a research institute, and a

---

[43] Articles VIII, VIII-A, and IX of the *Conservation Law* contain the provisions here summarized.

[44] Conservation Commission, *Sixth Annual Report* (1916), pp. 46-47.

bottling plant. The Conservation Department is frankly bullish on the State Reservation. "It must be assumed," it avers, "that the aim of New York State for The Saratoga Spa is not only to develop an institution of incalculable value to suffering humanity, but also to preserve the natural mineral waters for all time." [45]

New York's resources have been turned to the uses of recreation from the beginning of time. In particular, the State's forests, accentuated in the Forest Preserve, and its notable water resources, both lakes and streams, play important roles in the rapidly developing field of recreation.

New York's 78 state parks total almost 200,000 acres scattered from Long Island to the St. Lawrence and from Lake George to Niagara. The names themselves—Niagara, Thousand Islands, the Genesee River region, the Finger Lakes, Lake George—suggest the degree to which parks and park activities center on water resources. The three major water recreational activities are boating, bathing, and fishing. The State emphasizes each of the three. The Conservation Department maintains many boating facilities, and has published bulletins outlining canoe trips. The importance of the Barge Canal itself for recreational boating should not be overlooked, for thousands of pleasure boats ply the waters of the Canal from May to October. The 1956 report of the Division of Parks began with these words: "With a full six weeks of the best bathing weather in ten years, New York State Parks set new records in 1955." Park attendance for that year was well over 26,000,000 and revenue produced totalled almost $2,800,000. [46] Recreational activity in New York, which has been growing steadily during the last ten years, currently is booming. There is apparently no limit to the public's demand for and use of recreational facilities save that set by the economy itself.

Current Developments

Two emergent water programs worthy of special note concern irrigation and small watersheds. The farmers of New York have practiced irrigation on an individual basis for well over a hundred years, yet, so far as available records show, only 123 acres were under irrigation as late as 1899. The Water Supply Commission devoted three terse paragraphs to irrigation in its report of 1910. In the language reserved for new and strange things, it noted that "A most interesting

[45] Conservation Department, *Annual Report,* 1955, p. 233.
[46] *Ibid.,* pp. 179, 198.

example of irrigation is found at Irondequoit, near the city of Rochester, at which place water is obtained from underground sources and forced through pipes about the premises with a force pump operated by an electric motor, . . ." The Commission concluded that ". . . there is a large field in this State for the development of irrigation systems for the improvement of farm lands which now suffer greatly from drought in the summer season. . . ." [47] Twenty-five years later, for all this optimism, there were only 1,825 acres under irrigation.

Thereafter growth was rapid, with acres irrigated increasing more than three times from 1936 to 1939. It was not until after World War II, however, that agricultural irrigation in New York really hit its stride. As so often in the field of water practice, it was a technological development, this time aluminum tubing (which was light and easy to handle), that introduced a new period of rapid growth. Acres under irrigation almost doubled from 1939 to 1944, and again from 1944 to 1949; from 1949 to 1954 there was almost a three-time increase. The 1954 figure was slightly more than 59,000 acres, while the estimate for 1955 was 70,000. [48]

The Joint Legislative Committee on Natural Resources from its beginning took cognizance of the implications of irrigation for New York's agriculture. Each of its five reports (through 1956) dwelled on the subject, to which the Committee devoted considerable time and attention. The Legislature took a hand with the creation, in 1955, of a Temporary State Commission on Irrigation. The act creating the Commission cited the competitive disadvantage suffered by New York farmers through their failure to practice irrigation more widely, and charged the Commission ". . . to make a thorough study and investigation of the practicability of the use of irrigation projects as a means of expanding and stabilizing the agricultural economy of New York state. . . ."

The Joint Committee on Natural Resources and the Temporary Commission on Irrigation have concentrated their attention on two major areas where the potential for irrigation development appears greatest. The first of these is Long Island (and more particularly Suffolk County), where the practice of irrigation is already far advanced. Of the 65,000 acres under cultivation in Suffolk in 1955, 36,000 acres representing 620 farms were irrigated. [49] Water for the irrigation of Long

---

[47] Water Supply Commission, *Sixth Annual Report* (Albany, 1911), pp. 141-142.
[48] Temporary State Commission on Irrigation, *Report, 1956* (Legislative Document, 1956, No. 29), p. 21.
[49] *Idem., Report, 1957* (Legislative Document, 1957, No. 27), pp. 22 and 47.

Island farms is obtained from wells. The second major area comprises
the Ontario Plain from Rochester west to Lockport, an area of approximately 341,000 acres of which it has been said about 239,000
would respond well to irrigation. As late as 1950, only 19 farms total-
ling 600 acres were irrigated in this section. The 1955 figures, 84 farms
and 3,259 acres, represented a great increase percentage-wise, but the
acreage irrigated remains a very small fraction of the area which
would profit from irrigation.[50]

Important further developments in irrigation in New York await
the resolution of two difficult problems. The first, oddly enough in a
State always thought to have plenty of water, concerns water supply.
There are several areas where conditions appear ripe for the expanded
use of irrigation but where an adequate water supply is not available
at reasonable cost. The vegetable and small fruits-producing area south
of Buffalo is one such location. There the summertime flows of the
streams, already being used for irrigation to their maximum capacity,
provide only a fraction of the needed water. The condition in the lower
Hudson Valley is comparable. There farmers compete with municipalities for the use of summertime stream flows. Because of the interde-
pendence of town and country, adequate water for both must be
found if either is to develop soundly.

East of Rochester a special problem in agricultural water supply is
found. There the Barge Canal follows the natural drainage channel for
the Finger Lakes region. There is a large snap bean industry in this
area, and both processors and farmers are looking for extra water for
irrigation. The Genesee River through its connection with the Canal
at Rochester could provide the needed water, but unfortunately a big
new dam constructed on that stream by the Corps of Engineers (Mt.
Morris, 1952) was designed and is operated for flood control alone.
Its reservoir lies empty until the time of flood, when it fills (or partially
fills), only to be emptied again as rapidly as the channel below will
carry the surplus water. The Commission on Irrigation has noted this
condition, and has raised the question of at least limited multi-purpose
operation of the reservoir.[51] The resolution of this issue will be watched
with interest by all concerned with maximum utilization of the State's
water resources.

A second major problem confronting irrigation has to do with water
law, and particularly with water rights under the existing law. There
is no substantive statute on the subject of irrigation, and there is very

[50] *Ibid.*, p. 73.
[51] *Ibid.*, pp. 88-90.

little case law. Irrigation, which represents a new and potentially large demand for water, finds itself hampered by a system of law which took form before substantial agricultural need put in its appearance. This problem is discussed in some detail in Chapter IV, and will not be elaborated further here. The Commission on Irrigation has drafted two important "study bills" relating to irrigation, one concerning water supply for agricultural use, the other permitting the creation of county irrigation districts. These measures appear to be steps in the right direction, although it must be emphasized that the legal difficulties confronting irrigation in New York are complex and of long standing, and that their mitigation is likely to require sustained and patient effort.

The relations between water for agricultural use and flood control on the one hand and farm practices on the other have been recognized for many years. The State College of Agriculture has, of course, been active in the improvement of agricultural practice since its foundation. Its hand was greatly strengthened by the passage, in 1935, of the Federal Soil Conservation Act, which was followed in 1940 by an enabling act for New York State. Under these laws soil conservation districts have been established in 46 counties. The soil conservation movement is under the general aegis of the State Soil Conservation Committee, which, though a state agency, operates in the College of Agriculture. With the encouragement and assistance of soil conservation representatives, farmers for more than fifteen years have engaged in such water management practices as contour plowing, reforestation, construction of diversion ditches, stream bank protection, and construction of farm ponds. As a single example of the results of this activity, it may be mentioned that by 1956, 8,500 ponds had been completed with Soil Conservation Service help. These are used primarily for domestic and stock water supply, though their potential use for irrigation is not without significance.

Since its creation the Joint Legislative Committee on Natural Resources has been close to the soil conservation movement, and the State Commission on Irrigation has established a close relation with the movement in the last two years. The interest of the Federal government in the control of little waters has not been lost on these state agencies, which have followed developments in Washington with avid interest. Both recognized the potential significance of the Watershed Protection and Flood Prevention Act of 1954 (Public Law 566) for New York State, and both expressed satisfaction with the amendment of 1956. Acknowledging certain deficiencies in state law with respect to action contemplated by the Federal statute, the Chairman of the Joint Legislative Committee on Natural Resources introduced, and during its 1957 session the Legislature passed, a measure to bring New York organization and procedure into harmony with Federal requirements (Chap. 962, Laws of 1957). The act authorizes the creation in any county of a small watershed protection district; provides for the close association of such district with the county's soil conservation district (where one exists); permits the district to enter into local contractual relationships designed to effectuate its programs; authorizes the watershed districts of two or more adjoining counties to co-operate in the construction and operation of a joint small watershed project; and designates the State Commissioner of Agriculture and Markets as the agent to represent the State in negotiations concerning the small watershed movement. With the passage of this act, New York appears prepared to take full advantage of the Federal Watershed Protection and Flood Prevention Act.[52]

The small watershed movement looks to upstream flood control as its principal purpose. That the protection visualized will serve a number of auxiliary purposes is, however, not to be doubted. One secondary end is the prevention of siltation, another recreation, another irrigation, yet another water supply. Here, then, is a brand-new activity which may well develop into a multiple-purpose program of first importance.

INCENTIVES TO PROGRAM DEVELOPMENT

A program of public action normally may be supposed to grow from a felt need—that is, from a need experienced by a sufficiently influential and articulate public over a long enough time to produce action on the part of those who make decisions in the name of the government. Action programs result from policy decisions made by the political officers, which include the Governor and his immediate associates and the two houses of the Legislature. Program activities almost always find authorization in law. Ordinarily the legal basis of a given activity will be found in a statutory enactment, though it may be discovered in an administrative action or even in a judicial decision. The bulk of New York's water programs rests on statutory law, which means that a crucial decision point in respect to water programs is the Legislature.

The legislative body characteristically takes action either in response

---

[52] That that act vests little control in the hands of the State, while noteworthy, is not relevant to the present discussion.

to a persistent demand voiced over a long period, in reaction to a violent occurrence which creates an immediate and irresistible need, or in reflex action on a "sneaker" bill which has the good fortune to escape controversy. Three aspects of the problem of water supply, all developed at some length in the preceding section, will serve to illustrate the first two kinds of action.

That we may have either too much or too little water a good part of the time may be regarded almost as a basic law of water supply. The application of this law to two of the State's persistent problems will demonstrate how water programs come into existence in response to long-felt need. New York has some 300,000 acres of swamp, marsh, and bog lands which have too much water; at the same time it has several hundred thousand acres of farm land which at certain times have too little water—the Ontario Plain, to illustrate, which contains 239,000 acres of irrigable land, needs 20 inches of water during the growing season, but gets only 15 inches. These opposite conditions of too much water and too little have led to a demand for state assistance which in the case of drainage has persisted at low pressure for upward of a century, in the case of irrigation for a matter of no more than a decade. The Legislature has responded to the two demands by entering upon a drainage program and by creating a commission to study and to bring in recommendations with regard to the problem of irrigation. The response has not been spectacular because the need has not been desperate nor the demand violent.

Occasionally a need or a demand for programmatic action takes form as the result of a catastrophe. Here again two phases of the problem of water supply will serve as illustrations. The less violent takes the form of drought, which may be devastating over an extended period if not cataclysmic in its immediate consequences. New York City's water shortage of 1949 forced the City to re-double its efforts to increase its water resources. A more spectacular manifestation is found at the opposite extreme in flood, which frequently leads to immediate and dramatic public action. The Delaware Basin floods of the summer of 1955 and the spontaneous reaction they produced are of too recent occurrence to require elaboration. The great floods of 1935 and 1936 resulted in both Federal and state flood control statutes in a matter of months. Even though the importance of flood control had been officially recognized for a great many years, effective action was deferred until a catastrophic event riveted attention on the problem. The disaster produced an ironical turn in the environment in which a hearing was conducted by the Corps of Engineers Board for Rivers and Harbors on New York's protest against a recommended reduction in the flood control funds tentatively allocated to that State. The hearing was conducted in the War Department munitions building in Washington on the very day when flood waters from the rampaging Potomac River ran through the streets and lapped at the walls of the building itself. The hearing resulted in a recommended increase of 220 per cent over the amount previously found to be economically justified.[53] In earlier days, disaster frequently took the form of an epidemic which directed public attention sharply to the matter of water supply. New York City's epidemics of yellow fever and Asiatic cholera of the first part of the last century served as frantic stimuli to the search for new water. Again, in 1926, 15 per cent of the residents of the Village of Akron came down with typhoid fever and twenty-one of them died. The source of the outbreak was traced to the community's water well, which was adjacent to a creek. Following heavy rains, the creek rose to the point where it contaminated the well, with the result noted. After this outbreak, the Village abandoned the polluted well and developed an upland source of water, as it had been officially advised to do six years before. It is worthy of the record that the offending stream bears the name of Murder Creek.

Hand in hand with need, whether long-felt or catastrophic, goes vigorous leadership, for programmatic demands find their ways into action programs smoothed by energetic champions. It has been noted that DeWitt Clinton, more than any other one man, was responsible for the Erie Canal. A latter-day counterpart to Governor Clinton is found in Alfred E. Smith, who championed the cause of public development of the State's power resources from 1920 to 1930 and who in the end gained acceptance of his proposal for a State Power Authority. Programmatic leadership is by no means confined to the Executive Department, for legislative leaders, too, have made their influence felt. Currently several of the joint legislative committees have vigorous chairmen who pursue their duties with energy. The qualities of leadership also are found frequently at the second level of administration, where an executive secretary (or equivalent) often makes a distinct imprint on an agency program.

Public opinion plays a fundamental role in both the formation and the execution of government programs. The democratic basis for all public action is recognized in New York's programmatic laws, many of which specify that action may be taken only in response to a positive

[53] New York State Flood Control Commission, *Report, 1936* (Legislative Document, 1937, No. 59), pp. 16-17.

request by those involved. Thus water supply, drainage, and sewage disposal programs, where state action is involved, depend in the first instances on local initiative. The legislative committees and commissions active in the water resource field conduct hearings systematically on proposed new legislation of importance, as a matter both of remedying defects in the draft and of informing the people of the prospective change in program. The "study bill" device, by which a bill is given full publicity well in advance of its formal introduction in the Legislature, rests upon recognition of the importance of public opinion in a democracy. The leaders interested in a particular programmatic area rely heavily upon the support of public opinion in bringing demand for action effectively into focus.

Inextricably related to public opinion are the activities of pressure or interest groups, which even in so specialized a field as water resources appear in almost infinite variety. Professional, social, recreational, and fraternal organizations exist in such profusion today that no citizen who counts himself successful would care to admit that he was a member of less than half a dozen. Such organizations have (or may have) legitimate interests in public programs, and in any case they have the undeniable right to voice their opinion regarding any program proposal made. In the early days of this century the Merchant's Exchange of Buffalo maintained a Canal Bureau, which through public meetings and an energetic mail campaign supported the cause of the proposed barge canal. The Friends of the Forest Preserve have long resisted the invasion of the Forest Preserve by storage reservoirs; and in 1953 not less than 25 "forever wild" organizations joined with the Friends to write a prohibition on regulating reservoirs in the Forest Preserve into the Constitution (Chapter VI, below). On July 8, 1956, the *New York Times* carried a full-page advertisement under the caption, "Labor Fears Government Development of Niagara." The history of water resources in New York is replete with cases of programs both adopted and defeated by the action of pressure groups. The question of the legal and ethical aspects of such action is beyond the limit of the present discussion, which is intended only to suggest the importance of interest groups as a source of inspiration and support for (and opposition to) public water programs.

Social and economic change is a major incentive to the development of water programs. The rapid increase in population has resulted in grave problems of water supply. This is true particularly of the cities, which have been involved in an ever-widening search for new water sources for a century and a half. The more recent movement of population to the suburbs has produced its own peculiar problems and its own special solutions in such devices as water districts and inter-governmental arrangements for water supply. The rapid growth of industry also has led to serious problems both of supply and of pollution. The City of Syracuse recently yielded responsibility for supplying water to suburban industry to the Onondaga County Water Authority. The pollution of the Buffalo River by a cluster of industrial plants has led to a joint government-industry attack on the problem (Chapter VII). Two of New York State's most important water agencies, the Water Power and Control Commission and the Water Pollution Control Board, were established in good part to find solutions to the many and complex problems occasioned by social and economic change.

Related to social and economic change but worthy of separate mention is technological change. The history of the state canal system can be written largely in terms of technical developments in transportation and communication. It is ironical that so successful an enterprise as the Erie Canal should have been faced, almost at the moment of completion, by competition from the railroads, whose spectacular development completely changed the face of inland navigation between 1850 and 1900. On another front, research within the last very few years has demonstrated the value of irrigation of farm lands in the "humid east," while contemporaneous development in equipment has made irrigation practicable under the conditions which prevail in New York. So taken-for-granted a current phenomenon as air conditioning has resulted in the overtaxing of some urban water supplies, and in a move to regulate its use. On the other side of the ledger, modern purification methods have made possible the employment of water sources which would not have been reckoned safe a few short years ago. The development of good water from the sea, if and when a process is perfected that is both practicable and economical, will have a profound influence on the problem of water supply.

A major incentive to state water program development is found in local inability or unwillingness to cope with a problem. A wide variety of state programs have come into being because of local limitations of jurisdiction or resources; while a considerable percentage of the Legislature's time and thought is devoted to ways of making local programs more effective and local units more nearly adequate to deal with the problems confronting them. Contrariwise, the local community occasionally makes an important contribution to the State. To illustrate, in 1945 the City of Newburgh, in collaboration with the State Department of Health, began the fluoridation of its water supply on an experimental basis, a practice adopted as official state policy six years later.

Evidence of state adoption of water programs in reaction to Federal incitement is to be found on every hand. Passage of the Federal Power Act in 1920 was interpreted in New York as an invitation to pass a companion act; and the Legislature responded in 1921 by enacting a law defining a new state power policy. The Federal Flood Control Act of 1936 was matched by a State Flood Control Act passed the same year. The New York statute created a State Flood Control Commission to ". . . act as the agency of the State in assisting in the institution and consummation of a federal long range program of flood control. . . ." in the State. The Federal Soil Conservation Act (1935) was followed by a state law (1940) designed to facilitate action in pursuance of the national program; and the Federal Watershed Protection and Flood Prevention Act resulted almost immediately in a state act clearing the way for local participation in the small watershed program. Other illustrations of Federal incitement to state action exist in profusion, but these four will provide a basis for several comments on the subject. First, Federal encouragement and assistance frequently serve to energize a state program which otherwise might have continued to languish or which might never have come into being at all. Second, the Federal government through its stimulation of state action exercises a considerable influence on both the substantive programs and the administrative organizations of the states, and not alone in the water resource field. Illustrations of state programs and agencies which would not have existed or which would have taken other forms but for Federal subvention are many and readily available. Third, and as a corollary, Federal programs and agencies tend to beget their kind at the state level, with the result that national errors as well as national strengths are perpetuated in state practice. Fourth, one may wonder whether the State often rejects a Federal invitation to action which carries with it the promise of financial aid. Fifth, a Federally-inspired program may sometimes act as a depressant on one or more state programs. It appears a little far-fetched to expect that further work will be undertaken under the River Improvement and the River Regulating Acts, which require the beneficiaries to pay for the improvements, when Federal flood control structures can be had at comparatively little local cost. The net value of Federal contributions to state water programs is not in question here but only some of the collateral consequences of Federal-state collaboration.

Some Observations on Program Development

A number of conclusions regarding program development may be drawn from the New York experience in water-resource programs. Among these may be noted the fact that cataclysmic events, as war and depression, may either stimulate or depress state programs. The state canal system throughout its history has reacted sensitively to the alternate periods of "boom and bust." To cite two examples, the Delaware and Hudson Canal, which depended largely on the shipment of coal from the Pennsylvania fields to New York City, never did recover from the depression of 1857; and the Barge Canal, which suffered terribly from the doldrums of the depression, was ready nevertheless to accommodate the vast oil shipments into the interior which came after World War II. New York City's plan to develop the Delaware River as a source of water supply languished because of difficulties in securing financing during the depression. That hurdle cleared, work on the project began, only to be stopped during the war through inability to obtain materials. The construction of water supply and sewage disposal systems rose and fell with Federal policy during the depression. Both expanded in response to military and defense needs during the war and the years following. Basic water programs are peculiarly sensitive to the influence of catastrophic occurrences. They are essential to the maintenance of life itself, and must be protected at all costs; they possess the flexibility which makes possible rapid expansion, and so afford a cushion against the need for public works in time of depression; and they constitute a basic element in any wartime or defense construction program.

A second conclusion warranted by the data examined is that few programs once actively launched are ever abandoned. There is, of course, an occasional false start, since political leaders sometimes embrace a program demand which later turns out to have been spurious or ephemeral. Such occasions are rare, however, for programs normally come into being over a long period calculated to test well the genuineness of the need they purport to represent. Programs frequently change in emphasis, as from river improvement to river regulation to flood control; but a basic program has remarkable staying power both because the need was genuine in the first place and because it is continuing in nature. Another reason for program perpetuation is found in the crystallization of cliental (consumer or user) support behind the activity in question. Many feel that the Saratoga mineral springs development has outlived its usefulness as a state enterprise, but suggestions that it be jettisoned encounter a phalanx of defenders—led by senior citizens who hark back to the good old days—which thus far has proved invincible. Still another resides in the interests which soon come to vest in a particular program's bureaucracy. Both the political (commission or committee members) and the administrative (staff) bureaucracies quickly arrive at the conviction that the task they are engaged in is highly significant, and they set about to make it seem as important in the eyes of others as it is in their own. This accounts for the longevity of "temporary" commissions, for the tenacity and loyalty of administrative personnel, and, in measurable part, for program permanence. It also accounts for the continuation of an occasional program after the need it was designed to serve has disappeared.

The same factors which favor the perpetuation of a program work also for its extension, from which it may be concluded that an activity once undertaken may be expected to expand. There are, as before, two principal reasons for this phenomenon. First, the program gives satisfaction to real needs; but second, there are important inner forces working for its expansion. Administrative officials are prone to measure the importance of their programs by increase in activities, growth in budgets, and expansion in personnel. In government as elsewhere, nothing succeeds like success, and a program to be successful must appear prosperous. A program, along with the agency which symbolizes it, may progress (expand) or retrograde; but it cannot long stand still.

There is, of course, some contrary evidence on the subject of the tendencies toward program permanence and program expansion. A program occasionally withers away for want of support, either because the need was synthetic in the first instance or because the measures devised for its satisfaction were artificial or unworkable. Ordinarily the demise of a program may be attributed to precipitate action in adoption or to lack of ingenuity in implementation. In either case, the unusual spectacle of the attrition of a program would not seem to upset the overwhelming evidence that permanence and expansion are almost universal characteristics of government programs.

It must be noted that, as a program now and then atrophies for lack of support, a formally adopted program occasionally fails to get off the ground. The program outlined by the Fuller Act of 1907 (discussed above) comes readily to mind. That program dawdled along for a quarter of a century before its ultimate (and then only partial) activation in the Power Authority Act of 1931. Again, the River Improvement Law (1904) delineated a brave policy, but only one improvement project was ever launched under the act—and it concerned a creek rather than a river. In brief, there were no takers for these programs, which therefore failed to become operative.

It goes without saying that the climate of government strongly affects program development, yet the point is worthy of emphasis. The Fuller Act, which called for the development by the State of its water power resources, was passed at the instance of the able and resourceful (and very sober-minded) Governor Hughes; but it was never really accepted by the Legislature, which seized an opportune time (in 1915) to enact a contrary policy into law. The people of New York were individualistic, Republican, and private-business oriented in those days; and the state government reflected their predisposition, notwithstanding the frequent election of a Democrat to the governorship. It remained for the depression and the New Deal to pierce this armor with the various Federal aid programs. The Federal government has been expansionist from 1933 to this day, first by choice as a means of combatting the depression, then by force of circumstances beyond its control. New York likewise has joined in the course of history, wherefore the climate has favored an expanding state program for twenty-five years. New York, too, reflects welfare-state thinking, though not necessarily by conscious choice.

Last among the observations to be made on program development is the fact that new programs usually are placed in the hands of new agencies. In earlier days this was done forthrightly, with new boards and commissions created to handle new functions as the need seemed to arise. Since 1927, the Constitution has prohibited the creation of new departments, which now number nineteen and which are listed by name in the Constitution (Art. V, Sec. 2). Nowadays therefore a bit of wile is required, as in the establishment of the Water Power and Control Commission as the head of a division in the Conservation Department, and in the creation of the Water Pollution Control Board in the Department of Health. At the same time, there is a saving clause in the Constitution to the effect that the limitation on the number of departments ". . . shall not prevent the legislature from creating temporary commissions for special purposes . . ." (Art. V, Sec. 3). The Legislature has availed itself of this loophole more than once, as for example in the creation of the Temporary State Flood Control Commission, which has been continued on a year-to-year basis for twenty-two years. Thus while New York is not as free as most states to create new administrative agencies, there are ways to ensure that new programs will receive the special consideration thought to be due them.

Summary

The process by which a new activity is absorbed into the State's total program normally is long and involved. The initial request for action may result immediately in legislative response in the form of a statute; but in the case of an important program area it is more likely to lead, after considerable preliminary discussion, to the appointment of a joint legislative committee or a temporary commission to make studies and report back to the legislative body. Once the Legislature has determined to take positive action, the law passed may still be tentative and experimental, although it is less likely to be so if there has been careful anterior consideration of the problem. The evolution of a viable program—for water pollution control, for example—is a tedious and time-consuming process, extending frequently over several years and occasionally over half a century. And the most carefully conceived of programs is of course subject to modification, for a program must remain sufficiently flexible to meet new and changing needs.

It is not without significance that the title of this chapter is "Water Programs," for New York may not be said to have a water *program*. Its several programs have taken form over the course of many years, and each has evolved in response to the stimuli peculiar to its field. It is not improper to raise the question whether these eight or ten important individual programs add up to a satisfactory total program overall. There is evidence to suggest that they do not, if only because no single agency has ever been charged with responsibility for examining the whole problem of planning and managing the State's water resources. This is not the time to propose an answer, but it may be the place to raise the question whether New York might with profit undertake to review its total needs and to establish a concerted program designed to meet them. More will be said on this subject later.

Meanwhile, it is worthy of note that the decade divided by 1900 produced something approaching an administrative fault line. Before that time, the State's incipient water programs relied heavily on voluntary citizen cooperation, with the threat of court action in the background to effect compliance with programmatic regulations. Administrative organization, where found, was rudimentary. About 1900, thought began to turn to administrative structure, as was evidenced by the creation of the River Improvement Commission (1904) and the Water Supply Commission (1905). The establishment of these agencies signalized New Yorks' entrance into the administrative age, in which particularized government by statute and judicial decision yielded to generalized rules fixed by law but elaborated and executed by administrative agencies. We move now to an examination of the structure established by the State for the administration of its water-resource programs.

CHAPTER III

Administrative Organization

A PRELIMINARY review of the agencies involved in the administration of water resources programs in New York reveals a complex and varied structure. On the one hand, there are literally hundreds of private organizations active in the field. Their significance for water programs is great, but they are passed over here in favor of emphasis on public action. On the other, governmental agencies at every level take part in the management of water programs. It is the purpose here to center attention on these agencies, and more especially on New York's organization for water resources management.

The analysis must begin with a definition of terms, for every important word in the phrase "state administration of water resources" not only admits of a number of interpretations but is in fact used in a variety of senses. The distinction between federal and state affairs, while sometimes confused when the emphasis is on action taken, ordinarily is not difficult to make when the central focus is administrative machinery; but the distinction between state and local agencies is not always apparent, nor, when perceived, always easy to verbalize or to explain. Thus an occasional agency which appears to be local in character nevertheless must be classified as a state instrumentality by virtue of its legal status; the Saratoga Springs Authority is a case in point. Others which are quite similar on the surface deserve to be classified as local because of their origin and legal character; the Onondaga County Water Authority is an example. Similarly the Niagara Frontier Port Authority, whose New York directors are appointed by the Governor, is classified as a state agency, while the Oswego Port Authority, whose board members are appointed by the Mayor of Oswego, is excluded as a local agency.

The first criterion for inclusion or exclusion of a particular agency therefore turns on its legal status. It must be added that appearances often are deceiving, and that some decisions to include or exclude are open to question.

A second criterion hinges on the relations of the activities of a particular agency to water resources. For present purposes, a number of agencies which have only an indirect or a general subject matter relationship to water resources are omitted from consideration. Illustrative of such agencies is the Atlantic States Marine Fisheries Commission, whose tangential connection with water problems is apparent from its name, and the Department of Public Service, which discharges certain functions peripheral to water resources. Here again there is abundant room for interpretation and the element of judgment plays an important role, as before.

Whether or not a particular agency is administrative in character would seem to be a matter readily to be determined, but such unhappily is not always the case. The Temporary State Commission on Irrigation, for example, is characterized as a "study commission"; but its work is sufficiently important and sufficiently relevant to require mention in an analysis of administrative agencies. Two joint legislative committees deserve attention here; for although they are essentially legislative bodies, their work is very closely related to the administration of the State's water resources programs. Again, the State has membership on a number of interstate commissions, some of which are administrative in character, some not. These, too, seem to be worthy of inclusion. It will be clear from these observations that the term "administrative" has not been applied rigorously to exclude agencies whose activities are importantly related to water resources.

Two concluding comments of a general nature need to be made. First, not all state agencies involved in the administration of water resources programs are equally significant in the tasks they perform. Some, indeed, are relatively inactive, and some, though legally in existence, apparently are defunct in fact. For example, the current issue of *The New York Red Book* lists and describes the Committee on Water and Land Resources; but a member of that committee states that, so far as he knows, no meeting has been held since 1954. Some of the agencies listed here probably were never of much importance, some have atrophied through non-use, and some are of such recent origin that they have not yet been called actively into being.

Finally, it should be pointed out that it is virtually impossible to set up a system of mutually exclusive categories of state agencies involved in water resources programs. The agencies to be examined here are classified into six major types, but several require double listing. The Water Power and Control Commission, for example, is a commission, by title at least; but it is also head of the Division of Water Power and Control of the Department of Conservation. Similarly the Water Pollution Control Board, which must be listed as a board, operates within the Department of Health. The Port of New York Authority, here classified as a public authority, might well have been listed with the interstate commissions.

With these words of explanation and caution, we may now proceed to an analysis of the state agencies involved in the various water-resources programs. A classified list of these agencies follows. It is necessary to emphasize that another grouping almost certainly would differ from this in both arrangement and extent. Even so, the list at hand will serve satisfactorily as a framework on which to base a description of the State's administrative organization in the water resources field.

TABLE 2

State Water-Resource Agencies

I. *State Departments*
   A. Executive Department*
   B. Conservation*
      1. Division of Water Power and Control
         a. Water Power and Control Commission
         b. River Regulating Districts
      2. Division of Fish and Game
      3. Division of Parks
         a. State Council of Parks
         b. Individual State Park Commissions
      4. Division of Saratoga Springs Reservation
         a. Saratoga Springs Commission
         b. Saratoga Springs Authority
      5. Division of Lands and Forests
   C. Health*
      1. Water Pollution Control Board
      2. Bureau of Environmental Sanitation of the Division of Local Health Services
         a. Water Pollution Control Section
         b. Sewage and Waste Disposal Section
         c. Water Supply Section

  D. Public Works*
     1. Division of Construction
     2. Division of Operation and Maintenance
  E. Agriculture and Markets
  F. Commerce
  G. State

II. *Commissions*
  A. Water Power and Control Commission (see I.B.1.a. above)
  B. Saratoga Springs Commission (see I.B.4.a. above)
  C. Temporary State Flood Control Commission*
  D. Temporary State Commission on Irrigation*
  E. Albany Port District Commission
  F. Niagara Frontier Port Commission

III. *Public Authorities*
  A. The Port of New York Authority
  B. Power Authority of the State of New York*
  C. Northwestern New York Water Authority
  D. Niagara Frontier Port Authority
  E. Saratoga Springs Authority (see I.B.4.b. above)

IV. *Miscellaneous Agencies*
  A. Water Pollution Control Board (see I.C.1. above)
  B. State Soil Conservation Committee
  C. River Regulating Districts (see I.B.1.b. above)

 V. *Interstate Commissions on which New York is represented*
  A. Interstate Sanitation Commission
  B. Ohio River Valley Water Sanitation Commission
  C. New England Interstate Water Pollution Control Commission
  D. Interstate Commission on the Delaware River Basin
  E. Interstate Commission on the Lake Champlain Basin
  F. Delaware River Basin Advisory Committee

VI. *Joint Legislative Committees*
  A. Joint Legislative Committee on Interstate Cooperation
  B. Joint Legislative Committee on Natural Resources

If each state department be counted one unit (without reference to its sub-divisions), the list totals twenty-nine agencies. Of these, two certainly and perhaps three are legislative rather than administrative in nature, some might well have been excluded as local in character or peripheral in functions, and some are of comparatively little importance in programmatic terms. By any reasonable count, however, there are not fewer than twenty-five agencies responsibly engaged in the management of the State's water-resource programs. Administrative responsibility for the various statewide programs centers overwhelmingly on the agencies identified with an asterisk.

STATE DEPARTMENTS

A discussion of the State's administrative machinery for water programs properly begins with the Executive Department; for while this department has no responsibilities for water programs as such, the Governor by law performs a wide variety of acts which relate to the management of water resources. He appoints the heads of the state departments, the members of most commissions and the boards of directors of most authorities, and (some, but usually not all, of) the New York members of the interstate commissions on which the State is represented. His removal power likewise is great, and he exercises wide influence over the various water programs by reason of it. Many kinds of important actions become binding only with the Governor's signature: the proposed allocation of St. Lawrence power by the Power Authority (Chapter VIII) took effect only on his approval. The State operates under an executive budget prepared in behalf of the Governor by one of his principal appointees. The Governor of New York occupies a very strong position by reason both of tradition and of constitutional definition of duties and powers. He is the real as well as the nominal head of the State's government, from which it follows that he is also its principal officer in respect to its manifold water program.

The Conservation Department is chief among the state program departments in the scope and variety of its water activities. New York has been concerned about its natural resources since Colonial days, and measures were taken to protect certain resources (principally the forests) as early as 1700. It was more than 150 years later, however, before the State took action to establish an administrative agency for a resource program. The first such agency was the Fisheries Commission, which was set up by legislative act in 1868. Simultaneously support for a more vigorous public policy regarding the forests began to build up, and after many years of agitation the Forest Preserve Act was passed in 1885. The Act was highly significant in terms of forest policy, but it is remembered also for the administrative agency which it established. This agency, a three-man non-salaried Forest Commission, was the forerunner of the present Conservation Department. An indication of coming events is found in the combination of these two commissions in 1895, into the Forest, Fish, and Game Commission, like its principal predecessor an appointive three-man body.

Coincidental with the rise of public interest in fisheries and forests went a growing concern for the State's water resources. New York's interest in water transportation goes back, of course, to the early eighteen-hundreds, and there has been an administrative agency for the management of the canal system since 1826. Interest in the problem of water supply as such was not lacking, but it was not until toward the end of the last century that this took tangible form in a number of recommendations made by the Forest Commission between 1885 and 1902. The latter year saw the establishment of a Water Storage Commission which was directed to make a study of the whole subject of damage from floods and procedures for flood control. The report of this Commission resulted in the establishment of the River Improvement Commission in 1904. The new Commission was given the power to regulate stream flow, thus becoming the first general water control agency to be established by the State. The water was somewhat muddied by the creation, in 1905, of a Water Supply Commission; but it was cleared again (temporarily) when in 1906 the River Improvement Commission was abolished and its functions vested in the Water Supply Commission.

The first major step toward the consolidation of the several natural resources agencies was taken in 1911, when the Forest, Fish, and Game Commission and the Water Supply Commission were combined to form the Conservation Commission. Prior to 1911 the parent agency had operated alternately under a three-member body and a single commissioner. The Conservation Law, as re-codified that year, provided once more for a three-member commission, an arrangement which obtained until 1915. In that year the direction of the agency was vested in a single commissioner, and the basic organization has remained unchanged since.

Meanwhile, the problems of the production and sale of hydroelectric power became more and more pressing, and presently a demand arose for the creation of a special state agency to deal with the subject. This demand was met in 1921 with the almost simultaneous establishment of two Water Power Commissions, both made up of *ex officio* members. This anomaly was corrected in 1922 when the name of the second agency was changed to that of Water Control Commission.

The problem of consolidation was addressed once more in 1926 with the establishment of the Conservation Department in substitution for the Conservation Commission (Chap. 619, Laws of 1926). A year later the Water Power Commission and the Water Control Commission were combined into the Water Power and Control Commission, and the new agency was established in the Conservation Department as head of the Division of Water Power and Control. During the last 30 years the many new activities undertaken by the State in the conservation of natural resources as a general rule have been placed in the hands of the Conservation Department, whose basic organization has remained unchanged throughout that period.[1]

The functions of the Conservation Department include responsibility for managing state forests, waters, lands, and recreational facilities; custody of all wildlife resources; promotion of conservation education; administration of the State's water resources, including river regulation, licensing of water power sites, allocation of public water supply, and drainage; management of the state parks; and administration of the Saratoga Springs Reservation. In the discharge of its functions the Department operates through seven major divisions, as follows: Water Power and Control, Fish and Game, Parks, Saratoga Springs Reservation, Lands and Forests, Conservation Education, and Finance. The activities of the first five divisions are related more or less closely to water resources. Among these the Division of Water Power and Control is by all odds the most important to the present discussion.

As has been indicated, this division is headed, oddly enough, by the Water Power and Control Commission, which consists of the Conservation Commissioner (as Chairman), the Superintendent of Public Works, and the Attorney General, all serving *ex officio*. The chief deputies of these department heads may be designated to sit on the Commission as alternates for their principals, and this practice is generally followed.

The Division of Water Power and Control has a number of highly significant water management functions. The most important of these are worthy of note:

1. *Allocation of public water supply resources.* No new source of water supply may be acquired or used without the approval of the Commission. In particular, no new large-capacity wells may be put down on Long Island without Commission consent. The Commission, on receipt of properly executed local petitions, may itself construct, maintain, and operate reservoirs on state lands in the Forest Preserve as sources for municipal water supply systems. On receipt of notice of local action to form a union water district to supply water to two

---

[1] The basic reference on the history of conservation agencies in New York is Gurth Whipple, *Fifty Years of Conservation in New York State, 1885-1935* (Conservation Department and New York State College of Forestry, 1935).

or more municipalities, it may, if it finds the proposition a sound one, build and operate the facilities necessary to that end.

2. *River regulation.* This it may achieve through the establishment of river regulating districts, which in turn may construct dams and operate storage reservoirs. Two such districts have been created: the Black River Regulating District, which operates three reservoirs on tributaries to the Black River; and the Hudson River Regulating District, which manages one reservoir. Each district in turn has its own governing board of three members, all appointed by the Governor.

3. *Drainage of low-lying and swamp lands.* This may be accomplished either through a procedure which enables individual landowners to drain their land with state technical assistance, or through creation of drainage improvement districts. Five such districts have been set up; each is managed by the Commission direct, without local administrative organization.

4. *Management of the State's water power resources.* The Commission has the responsibility for issuing licenses for the development and use of water power sites (including both water and land) in which the State has proprietary rights and interests. It also has the duty of fixing and collecting rentals for state water used for the generation of power, as for example at Niagara Falls. The influence of the Governor is seen in the statutory provision that no license issued by the Commission shall be effective unless and until it is approved in writing by him.

The organization and powers of the Division of Water Power and Control, and the rules governing its procedures as well, are set forth in considerable detail in the Conservation Law.[2] The procedures specified vest the Commission with the power of subpoena, and with broad investigatory powers as well. They provide also that hearings shall be conducted with regard to any important action proposed to be taken, with full right to be heard guaranteed to all interested parties. They specify further that the Commission shall have the power to bring suits in the performance of its duties. Finally, they guarantee to any interested party the customary rights of legal recourse against Commission action.[3]

The Division of Fish and Game administers all laws relating to the State's wildlife resources. In particular, it issues hunting, fishing, and trapping licenses. It also operates six game farms and twenty-three

[2] *McKinney's Consolidated Laws of New York, Annotated,* Book 10, *Conservation Law,* Articles V-XIV.
[3] *Ibid.,* Sections 400-402.

game management areas, along with twenty-two fish hatcheries and a laboratory for the study of fish diseases. Of direct relevance to the present study is the Division's responsibility for conducting biological surveys of and developing fish stocking and management policies for the State's inland waters. The Division of Fish and Game is headed by a career man who is subject to civil service regulations.

The Division of Parks manages 78 state parks, many of which center on lakes and streams. Its principal emphasis is on recreation. The Division is administered by a Director of State Parks, who like his opposite number in the Division of Fish and Game is a career man in the state civil service. The State has ten park regions, nine of which are under the supervision of as many park commissions whose members, with a few exceptions, are appointed by the Governor; the tenth, including the Forest Preserve, is administered by the Division of Lands and Forests. The chairmen of the nine regional park commissions, together with the Director of Lands and Forests, comprise a State Council of Parks, which serves in an advisory capacity to the Director of State Parks.

In 1930 a temporary commission was appointed by the Governor to develop the Saratoga Springs Reservation as a state health resort. In 1937 the commission was incorporated in the Conservation Department, where it has remained since. Composed of twelve members appointed by the Governor, the Commission serves as head of the Division of Saratoga Springs Reservation. The functions of the Division center on the development and management of the Saratoga Springs properties for the public benefit. In the pursuit of this goal, the State in 1933 created the Saratoga Springs Authority to obtain a loan from the Reconstruction Finance Corporation for completion of the Saratoga Spa, which was already in process of development with state funds. Under an interesting interlocking arrangement the Commission, which on the one hand serves as head of the Division, on the other serves also as the Board of Directors of the Authority, and the chairman of the Commission is president of the Board.

As its title indicates, the principal functions of the Division of Lands and Forests have to do primarily with areas other than water administration. Nevertheless the Division bears certain responsibilities which have direct significance for water resources. Chief among these are its duties as administrator of the Forest Preserve, in which a number of streams have their headwaters. The management of the Preserve is important for its contribution to watershed protection, which is of continuing concern to the Division. Another aspect of its work which is related to water resources has to do with recreation in the Forest Preserve, responsibility for which is lodged in the Division. The Director of Lands and Forests, the administrative head of the Division, is a career person under civil service tenure.

The concern of the Department of Health in the field of water resources centers on water quality, and so on the problem of pollution. The laws of the State have long given the Department broad powers of supervision over the public supplies of potable waters; regulation of the discharge of sewage into public waters; and approval of realty subdivisions, whose developers were required to submit a plan for furnishing the subdivision with satisfactory water supply and sewerage facilities. The Water Pollution Control Act of 1949 brought these long-existent requirements into somewhat sharper focus and provided for a more vigorous means of enforcement.[4] The Act was careful to state that it was designed to provide "additional and cumulative remedies to abate the pollution of the waters of the State," and was not meant to abridge or repeal any laws on the subject then in existence.

The Department of Health is headed by a Commissioner appointed by the Governor. At the Commissioner's right hand is the Public Health Council, a body of nine members of whom the Governor appoints eight, the Commissioner serving *ex officio* as the ninth member. The Council has the power to establish, and to amend and repeal, sanitary regulations known as the state sanitary code, subject to approval by the Commissioner of Health. It also serves as an advisory body to the Commissioner, and may make such recommendations as it may deem appropriate.

The Department is organized into five major divisions, of which only one has important functions in the water resources field. The Division of Local Health Services has a Bureau of Environmental Sanitation, which in turn has a Water Pollution Control Section, a Sewage and Wastes Section, and a Water Supply Section. It is in these three sections that the Department's action programs in water management center. They have an organic relation to the Department's six regional offices and fifteen district offices, which in turn maintain close relations with the sixteen full-time county health departments and the ten full-time city health departments. The Bureau of Environmental Sanitation thus has direct access to the Department's field organization and to local departments of health as well.

The Act of 1949 not only strengthened New York's water pollution law; it also created a Water Pollution Control Board. Established

---

[4] Chap. 666, Laws of 1949.

within the Department of Health, the Board in effect has become the Department's official arm for the administration of its water program. It would not be correct to say that it has superseded the Bureau of Environmental Sanitation, which has important duties other than those relating to water resources; but it is true that the Board has become the agency through which the work of two of the Bureau's three water-oriented sections is directed.

The 1949 law had two major purposes: first, to prevent new pollution; second, to effect the abatement of existing pollution. In pursuance of the first function, the law requires that, with respect to all new proposals for discharging polluting substances into the State's waters, plans for disposal systems shall be submitted to the Water Pollution Control Board, whose approval must be secured before construction may proceed. These requirements also apply to proposals to modify existing sewage or waste disposal systems. In the discharge of this function, the Board operates through the Sewage and Waste Disposal Section.

In pursuance of the goal of pollution abatement, the Board has set about systematically to classify the waters of the state into seven classes, ranging from AA (water fit for drinking with minimum treatment) to F (water suitable only for waste disposal). The Board operates through the Water Pollution Control Section in pursuing its work of pollution abatement. Proceeding basin by basin, it has made substantial progress toward its goal of classifying all the waters of the state.

The Water Pollution Control Board consists of the Commissioners of Health (as Chairman), Conservation, Agriculture and Markets, and Commerce, and the Superintendent of Public Works. As is customary with such agencies, the legally designated members may name deputies to serve in their stead. The Board maintains a permanent professional staff under the direction of an Executive Secretary.

The Department of Public Works is headed by a Superintendent of Public Works appointed by the Governor. The Department is organized into four major divisions, two of which, Construction and Operation-and-Maintenance, have important functions with regard to water resources. The Division of Construction is responsible for the construction work of all state departments and agencies, although there is an occasional exception, as the State Power Authority, which is authorized to undertake construction activities on its own account. The Division of Construction designs plans, prepares specifications and cost estimates, and undertakes actual construction of all structures having to do with canals, waterways, beach protection, and flood control and erected under state authority. Small projects may be handled by departmental personnel; the larger ones are executed by private firms under contract. The subdivision in charge of this work is headed by a deputy chief engineer.

The Division of Operation and Maintenance has three major subdivisions, one of which, under the direction of an assistant superintendent, is responsible for canals, waterways, and flood control. This subdivision maintains and operates the state canal system and all related reservoirs and feeders. It has charge of repairs of locks, dams, and bridges, bank protection work, snagging and debris removal operations, channel clearance and dredging, maintenance of drainage slopes, and repair of canal equipment and machinery. Further, it registers all pleasure boats and all public vessels which use the State's waterways, issues licenses for officers of public vessels, annually inspects all public vessels on intrastate waters, and issues permits for use of all canal facilities. Finally, it maintains and operates all flood control works and facilities.

Both the Division of Construction and the Division of Operation and Maintenance operate through the offices of the ten public works districts into which the State is divided. Each district office is in charge of a district engineer, and each is set up to reflect the basic organizational structure of the central office. The district engineers actually carry out the functions in their regions for which the division heads assume general responsibility at the state level.

The activities of the Department of Agriculture and Markets in the field of water management are neither many nor, at this time, important. The general interest of the Department in the subject has been recognized through inclusion of the Commissioner of Agriculture and Markets in such bodies as the Water Pollution Control Board and the State Soil Conservation Committee. A new function recently was vested in the Commissioner which in all likelihood will take on added significance in the future. Projects for which application for Federal aid is made under the Watershed and Flood Prevention Act (P.L. 566) must first be approved by the states. In New York, the Governor has designated the Commissioner of Agriculture and Markets to review and take action on such projects. It is too early to say what this designation may mean in terms of more active involvement in water management, though there are signs that the "small watershed" movement is gaining strength rapidly in the State.

The Department of Commerce may not be classed as a water resource agency in any except a very limited sense. Its basic function is to promote the economic well-being of the State, and to that end to encourage economic development, to confer with business and industry interests regarding economic opportunities, and to conduct studies relevant to these major purposes. In the discharge of its responsibilities, it sometimes gets into the water resources field, as, for example, when its economic surveys lead it into a study of water quality. Such instances are not unusual, but they are sufficiently rare to indicate that the principal concern of the Department of Commerce lies elsewhere than in the field of water management.

Like the Department of Commerce, the Department of State has only limited functions in the water administration field, though on occasion these can be important. Such functions vest largely in the Division of the Land Office, which is headed by an *ex officio* board consisting of the Secretary of State (as Chairman), the Attorney General, and the Superintendent of Public Works. The Land Office has responsibility for all state-owned lands not set aside for particular uses, and its activities may have significant effects for water resources.

COMMISSIONS

A commission (New York style) has been defined as an agency of the State created by statute to carry out a specific task. Some commissions are permanent, in that no limit is fixed for their life. Others are created for one year, although they may be continued on a year-to-year basis. The latter are called temporary commissions. The distinction is illusory; for an occasional "permanent" commission either never comes into being in the first place or is allowed to lapse for want of interest or support after a year or so, while now and then a "temporary" commission catches on and continues indefinitely by annual re-authorization by the legislature. Characteristically a commission will include members appointed by the Governor, senators appointed by the Temporary President of the Senate, and assemblymen named by the Speaker. Some commissions have *ex officio* members in addition. A recent count revealed approximately twenty-five such bodies. Of these, as many as six have functions which place them in the water resources field.

Two of these, the Water Power and Control Commission and the Saratoga Springs Commission, were mentioned in connection with the appropriate state departments. A third, the Temporary State Flood Control Commission, is an excellent example of a temporary commission which has become permanent by prescriptive right. Established in 1936, it has been continued by year-to-year enactment until almost nobody bothers to include the word "temporary" in its title. The agency's annual report appears forthrightly as the "Report of the New York State Flood Control Commission."

The Commission has twelve members, including four appointed by the Governor and four senators and four assemblymen named as above indicated. It works through a full-time Executive Director. New York of course participates in the Federal flood control program; the Flood Control Commission is one of the two principal agencies (the Department of Public Works is the other) for such participation. The major responsibility of the Commission is to represent the State in negotiations with the proper Federal agencies for maximizing flood control work in the State.

Established as recently as 1955, the Temporary Commission on Irrigation consists of eleven members, nine representing the Governor, the Senate, and the Assembly, and two (the Commissioner of Agriculture and Markets and the Attorney General) serving *ex officio*. The measure creating the Commission noted the need for a planned program which would enable New York farmers to compete with those of other parts of the country where irrigation is in use; and it directed the Commission to make studies of irrigation as a means of improving and stabilizing the State's agricultural economy. Originally created for one year with an appropriation of $25,000, the Commission in 1956 was continued for an additional year with an appropriation of $40,000 (Chap. 696, Laws of 1955; Chap. 706, Laws of 1956). It was continued for yet another year in 1957 (Chap. 380, Laws of 1957). If anything is to be learned from the history of the Flood Control Commission, a bright future awaits the Temporary State Commission on Irrigation.

The Albany Port District Commission was created in 1925 (Chap. 192, Laws of 1925) as a means of meeting certain requirements laid by the Federal government as a condition to its agreement to dredge a channel from New York City to Albany. The District embraces the cities of Albany and Rensselaer, together with limited adjacent territory. It is governed by a Commission of five members appointed by the Governor, four on nomination by the Mayor of Albany, one on nomination by the Mayor of Rensselaer. The Commission is empowered to prepare and adopt a comprehensive plan of port development, and, with the approval of the Corps of Engineers of the U.S. Army, to construct and operate the facilities called for by the plan. To this end, it is given broad powers to acquire, hold, and dispose of property. It is authorized further to fix and collect rates, rents, and fees for the use of all facilities owned by the District. It is also empowered to issue bonds. Finally, it is authorized to apportion any annual deficit between the Cities of Albany and Rensselaer, on the basis of the proportional benefits enjoyed by the two cities from the operation of the port. The cities in turn are directed to levy property taxes sufficient both to cover the annual deficits and to support the bonds issued by the Commission. The powers of the Albany Port District Commission thus are quite broad, and its legal position, if somewhat anomalous, is nonetheless a strong one.

By contrast, the Niagara Frontier Port Commission is nothing more than a paper organization. Created in 1951 (Chap. 802, Laws of 1951), the Commission was conceived as a joint agency of such of the cities and towns around the City of Buffalo as might elect to participate with that city in developing comprehensive port facilities. The law provided that the Commission should be made up of three residents of Buffalo and one from each municipality choosing to participate, all to be appointed by the Governor on nomination of the participating municipalities. In the end the municipalities decided not to collaborate, hence the Niagara Frontier Port Commission never actually came into being. Here clearly is an agency for which need was sensed but for which active support failed to materialize when the time came for action.

PUBLIC AUTHORITIES

In its exhaustive study of the subject, the Temporary State Commission on Coordination of State Activities found that, to 1955, New York had created fifty-three public authorities (exclusive of housing authorities), of which thirty-three were then active.[5] Authorities have been created to build and operate bridges, tunnels, highways, water supply systems, sewage disposal facilities, airports, ports, terminals, urban transit systems, parks, and parking lots, to name a few purposes at random from a long list. They are governed usually by a board of directors of three to five members (although some are larger), appointed by the Governor (normally) for terms varying from three to six years. The board of directors may consist of full-time salaried individuals or of members serving part-time without pay. In either case,

---

[5] Temporary State Commission on Coordination of State Activities, *Staff Report on Public Authorities under New York State* (Legislative Document, 1956, No. 46), pp. 4 ff. This undoubtedly will become the standard reference work on the public authority in New York.

the board of an active authority usually conducts its business through a full-time, salaried executive director, who in some instances is supported by a sizable staff.

Six public authorities have functions which relate them to the field of water resources. The oldest and by far the best known of these is the Port of New York Authority, which was created by an interstate compact between the States of New Jersey and New York in 1921. The Authority is managed by a Commission of twelve, with six appointed by the Governor of New York and six by the Governor of New Jersey. The Commission operates through a full-time Executive Director. The Authority is self-supporting, issuing its own bonds which it services from earnings. Among its many spectacular activities are those of constructing and operating port properties and in general managing the vast harbor facilities of the New York-Newark-Hoboken area.

The Power Authority of the State of New York was created in 1931 (Chap. 772, Laws of 1931). Its purpose was to represent the State in furthering commerce and navigation on the St. Lawrence, and more particularly to develop the hydroelectric power latent in the international rapids section of the river and to that end to construct and operate all necessary facilities, all in cooperation with the proper Canadian authorities. In 1951 the jurisdiction of the Authority was extended to cover the power made available from the Niagara River by the 1950 Treaty with Canada.

For more than twenty years the Power Authority sought ways to effectuate its mission, then spectacular things began to happen. In 1953 it was granted a license by the Federal Power Commission to construct and operate the St. Lawrence hydroelectric power project, in conjunction with Canadian representatives; and with this recognition the Authority took on new life. It has been much in the news of late by virtue of its vigorous prosecution of the St. Lawrence enterprise, and there is reason to expect that it will become even better known with the prospective development of the power available at Niagara Falls. Here is an arresting illustration of an agency which, as events proved, was created prematurely, but which conducted a holding operation in the intervening years and was ready with an established procedure and organization when the strategic time for action arrived.

The Board of Trustees of the Power Authority consists of five salaried members appointed by the Governor. It operates through a full-time General Manager and Secretary. The Authority finances its operations through sale of revenue bonds.

The problems, both intrastate and international, peculiar to the "Niagara frontier" have given rise to a number of special districts, commissions, and authorities with a considerable variety of purposes. Two of those currently in existence (or just emerging) take the form of public authorities whose activities center on water resources. The first of these, the Northwestern New York Water Authority, grew from the recommendation of a temporary state commission created to make a study of the water supply problems of the several counties extending eastward from the Niagara River. Acting on the commission's recommendation, the Legislature in 1950 (Chap. 806, Laws of 1950), established the Northwestern New York Water Authority District (to comprise all of five counties and parts of two others) and, coincidentally, the Northwestern New York Water Authority. In essence, the Authority was created for the purpose of developing an adequate supply of good water and distributing it throughout the district. The Authority was required to be self-financing, although the Legislature appropriated $50,000 in 1950 and another $50,000 in 1951 to help it get under way. As late as 1956, the Authority was still working on the problem of capital financing and had not yet begun operations. The Board of Directors consists of five part-time, unsalaried residents of the district appointed by the Governor.

The second of the two western authorities was created in anticipation of the good times expected to follow in the wake of the St. Lawrence Seaway development. The Niagara Frontier Port Authority, established with an initial appropriation of $75,000 by an act passed in 1955 (Chap. 870, Laws of 1955), is designed to improve the port facilities in the Buffalo area. There is evidence that those supporting this Authority have in mind the precedent set for a broad-scale development by the Port of New York Authority. There is a flaw in the Niagara plan, perhaps not a minor one: to provide a substantial revenue base, it is intended that the Authority absorb the Buffalo and Fort Erie Public Bridge Authority, which operates the Peace Bridge across the Niagara River; and this requires the agreement of the Canadian government. At the end of 1956 the necessary approval had not been secured, and the Niagara Frontier Port Authority was therefore without the major means of financing visualized for it.

MISCELLANEOUS AGENCIES

Among the miscellaneous state agencies, only one requires special mention. The State Soil Conservation Committee was established in

1940 in the State College of Agriculture as a state agency (Chap. 727, Laws of 1940, as amended throughout by Chap. 883, Laws of 1945). The law vests the Committee with important duties concerning soil conservation districts (which have now been established in as many as 46 counties); for present purposes, however, attention will be centered on the declaration of policy stated by the act rather than on the functions of the Committee. The law declares it to be legislative policy to provide for the conservation of the soil resources of the State, the control of soil erosion, and the preservation of natural resources; to contribute to the control of floods and the drainage of agricultural lands; to prevent impairment of dams and reservoirs; and to assist in maintaining the navigability of rivers and harbors. The dedication of the Soil Conservation Committee to these ends provides another administrative link between agricultural practice and water resource management. It is altogether likely that the Committee's activities in the water resource field will increase greatly in importance as the State moves into the inviting area of small watershed development. The Committee consists of ten members, five appointed by the Governor, five (including the State Conservation Commissioner and the Commissioner of Agriculture and Markets) serving *ex officio*.

INTERSTATE COMMISSIONS

In addition to the intrastate administrative (and quasi-administrative) agencies described to this point, New York maintains membership in a number of interstate commissions which treat problems beyond the State's jurisdiction. Some of these agencies rest on interstate compacts, some draw their authority from parallel state statutes, and some have no legal base; some enjoy modest financial support (normally through joint state contributions), some have little or none. They deal with a wide variety of subjects. As many as five pursue programs which are oriented toward water, and in each of these instances the emphasis is on pollution control.

New York participation in the first of these, the Interstate Sanitation Commission, was authorized in 1936 (Chaps. 3 and 4, Laws of 1936), and shortly thereafter a compact was entered into with the State of New Jersey. Connecticut became a party to the compact in 1941. The compact creates an Interstate Sanitation District centering on New York harbor but extending northward along the Hudson River for some miles and eastward along Long Island Sound to New Haven. It is the principal duty of the Interstate Sanitation Commission (the District's administrative agency) to control future pollution and to abate existing pollution in the adjacent tidal and coastal waters of the three signatory states. The Commission consists of fifteen members, of which five represent each State. Four of New York's commissioners are appointed by the Governor; the fifth, the State Commissioner of Health, serves *ex officio*. The Commission has a Director and Chief Engineer as its administrative officer.

Agitation for the control of pollution in the Ohio River came to a head in the late 'thirties in a proposed compact to be entered into by the several states concerned. New York's decision to participate was declared in 1939 (Chap. 776, repealed and re-enacted by Chap. 945, Laws of 1939); but it was nine years later before the last potential signatory state was satisfied and the compact duly completed and approved. The compact, ultimately agreed to by eight states, provides for an Ohio River Valley Water Sanitation District and for a Commission of three members from each state to carry into effect the declared policy of the compact, which is to abate and control present and future water pollution in the Ohio River drainage basin. Two of New York's three commissioners are appointed by the Governor; the third, the State Commissioner of Health, serves *ex officio*.

New York became a member of the New England Interstate Water Pollution Control Compact, to which all New England states save Maine are signatories, in 1949 (Chap. 764, Laws of 1949). The principal declared purpose of the compact is to establish an interstate commission to work toward the abatement of existing pollution and the control of future pollution of all interstate waters of the region. The Commission comprises five members from each of the signatory states. New York's five commissioners include four appointed by the Governor and the State Commissioner of Health, who serves *ex officio*.

Each of the three commissions thus far noted rests upon an interstate compact duly negotiated, agreed to by the signatory states, and approved by Congress, in harmony with constitutional requirements. The Interstate Commission on the Delaware River Basin is an agency of a different kind, for it depends for legal status on parallel legislation enacted in each of the four participating states. It is therefore more truly characterized as a joint state agency than as an interstate commission. New York became a party to the Incodel arrangement in 1936, with its participation formalized by legislative action in 1939 (Chap. 600, Laws of 1939). The Commission is made up of five members from each of the participating states. The New York members are designated by the Joint Legislative Committee on Interstate Cooperation. Incodel's announced program is a broad one, but its major achievements have been in the field of pollution abatement. The Commission maintains a small permanent staff under the direction of an Executive Secretary.

The Interstate Commission on the Lake Champlain Basin held its first annual meeting in 1939. It had no legal standing, no official representatives, and, worst of all, no appropriation. It consisted simply of unofficial delegates from New York and Vermont who came together each year to discuss problems of mutual concern arising from joint use of or interest in the Lake Champlain Basin. The Commission nevertheless was able to make good progress through voluntary cooperation in such areas as pollution abatement, wharf improvements, fish and game matters, and revived consideration of the improvement of the Hudson-Lake Champlain waterway.[6] The New York Joint Legislative Committee on Interstate Cooperation in 1956 expressed the view that the Lake Champlain Commission had earned its spurs, and recommended that it be given official sanction, and more particularly that it be voted funds to carry on its work.[7] Even as the Committee spoke, the Commission made the grade (so far as New York is concerned) through passage of an act designating it as that State's representative in planning the 350th anniversary celebration of the discovery of Lake Champlain (Chap. 876, Laws of 1956). There are nine New York representatives on the Commission, including the State Commissioners of Agriculture and Markets, Commerce, Conservation, and Health, all *ex officio*; the Chairman of the Joint Legislative Committee on Interstate Cooperation, *ex officio*; and four members appointed by the Chairman of the Joint Legislative Committee. The measure included a statement of more than temporary significance in its declaration of New York's policy ". . . to join with the state of Vermont in a common effort to promote the sound development of the underdeveloped resources of the Lake Champlain basin" (Sec. 1). It also appropriated $20,000 toward the expenses of the Commission, which, with statutory recognition and its own appropriation, may be supposed to have its foot firmly planted on the second rung of the ladder leading to success.

The Delaware River Basin Advisory Committee is not properly considered an interstate commission, but an informal committee instead. Established in 1956, it comprises six members appointed as personal representatives by the Governors of Delaware, New Jersey, New York, and Pennsylvania and the Mayors of Philadelphia and New

[6] Joint Legislative Committee on Interstate Cooperation, *Report* (Legislative Document, 1956, No. 66), pp. 184 ff.

[7] *Ibid.*, p. 177.

York City. Its original mission was to maintain close contact with and where appropriate to cooperate in the survey of the Delaware River Valley then in progress under the leadership of the U.S. Army Corps of Engineers. Subsequently the Committee established a research corporation which, with the support of a substantial grant from the Ford Foundation, sponsored a study of the problems of administering a broad water program for the Delaware Basin. The research corporation (with its own funds) also carries on a program of public education regarding the water problems of the Delaware Valley.

JOINT LEGISLATIVE COMMITTEES

The concern of the State Legislature for water resources problems is manifested in a variety of ways. Periodically it passes substantive acts in the field, and makes appropriations to water resources agencies. Its continuing interest is evidenced by its standing committees, several of which in each house have to do with water problems. Both Assembly and Senate have their Committees on Agriculture, Commerce and Navigation, and Conservation, and in addition the Assembly has a Committee on Canals and Waterways. Further, the two bodies maintain a number of study groups constituted as joint committees, of which at least three concern themselves with water problems. These are the Joint Legislative Committees on Revision of the Conservation Law, on Interstate Cooperation, and on Natural Resources. The last two are deserving of brief discussion here.

The Joint Legislative Committee on Interstate Cooperation, created in 1935, reflected New York's interest in the movement for interstate cooperation then gathering momentum. The resolution creating the Committee sang the praises of voluntary cooperation among the states, and declared it to be state policy to cooperate with the American Legislators' Association. The first report of the Committee (1936) dealt with a wide variety of subjects; but it gave evidence of a lively interest in water problems by listing, under "Future Work in Interstate Cooperation," conservation, planning and water resources, water pollution, flood control, and water supply.[8]

How well the Committee has fulfilled its early promise may be judged from an analysis of its 1956 report (Legislative Document, 1956, No. 66). There it is revealed that two of the six subcommittees deal respectively with conservation and water resources. The report of

[8] *Report of the Joint Legislative Committee on Interstate Cooperation* (Legislative Document, 1936, No. 111), pp. 129 ff.

the water resources subcommittee is especially significant as revealing a continuing and very lively interest in interstate water problems, particularly the problem of pollution.

The Joint Legislative Committee on Interstate Cooperation, as constituted by the 1955 concurrent resolution continuing the Committee for 1956, consists of seven assemblymen, five senators, and five state officials (as advisory members) designated by the Governor. The condition of the Committee is indicated by the same resolution, which was generous in its praise of the Committee and its work, and by the fact that the legislative appropriation for the Committee increased from $15,000 in 1935 to $35,000 in 1956. Clearly the Committee has established an important place for itself in the machinery of legislation.

Until 1951, the problems relating to the preservation and management of the State's natural resources were left to the Joint Legislative Committee on Interstate Cooperation. In that year the Legislature, recognizing the emergency resulting from the damage wrought (particularly to the forests) by the hurricane of the preceding year, by concurrent resolution created a Joint Legislative Committee on Natural Resources. The resolution charged the Committee to make a study of the State's natural resources, including ". . . its fish and game, its waters and the abatement of pollution therein, and the recreational and other uses appertaining thereto; . . ."[9] The Committee immediately concentrated on water resources as one of its principal areas of interest, and its first report dwelled at length on various aspects of that subject. The Committee's 1956 report revealed its continuing interest in the field.[10] Of the three special advisory committees to the Joint Committee, one dealt specifically with water resources and water rights. Two of the seven sections of the report dealt with water resources and water pollution, and the others related directly or indirectly to that subject.

The Joint Legislative Committee on Natural Resources consists of four senators, three assemblymen, and three advisory members to be appointed by the Governor of the State. The place which the Committee has made for itself in the esteem of the Legislature is suggested by the facts that it has been continued from year to year, and that its appropriation increased from $20,000 in 1951 (its first year) to $40,000 in 1956.

[9] *Report of the Joint Legislative Committee on Natural Resources* (Legislative Document, 1952, No. 77), pp. 13-14.

[10] *Report of the Joint Legislative Committee on Natural Resources* (Legislative Document, 1956, No. 63).

FEDERAL AND LOCAL AGENCIES

It would, of course, be wholly erroneous to suppose that the elaborate state structure for water management here described stands alone in the field. On the contrary, a number of Federal agencies are involved in New York's water programs. Chief among these are the Corps of Engineers of the U.S. Army, the Geological Survey (Department of the Interior), the Public Health Service (Department of Health, Education, and Welfare), the Forest Service and the Soil Conservation Service (Department of Agriculture), the Weather Bureau (Department of Commerce), and the Federal Power Commission. At the other extreme, there are literally hundreds of local agencies engaged in various aspects of water management. These are represented by municipal and other local water supply systems, county and city health departments, soil conservation districts, small watershed districts, and drainage and improvement districts in some variety. The purpose in noting Federal and local participation in water management programs at this point is merely to call attention to the fact that the state water agencies by no means "go it alone." It will not be necessary to describe these agencies or to analyze their operations, except to observe by example in the chapters to follow how they contribute to the work of the state agencies.

CONCLUDING OBSERVATIONS

It is clear that New York's machinery for the administration of its water resource programs is both complicated and highly fragmented. Considerable effort has been made to introduce a measure of consistency into the actions taken by the various agencies through *ex officio* and overlapping memberships. The Water Power and Control Commission and the Water Pollution Control Board, to cite two of the major water resources agencies, are composed exclusively of state department heads serving *ex officio*, while duplicating memberships are found in profusion. A principal consequence of this practice may be seen from the fact that, among 10 agencies identified as having a heavy inter-departmental flavor, one individual (the Commissioner of Health) serves as a member of five different boards, three (the Commissioners of Agriculture and Markets, Conservation, and Commerce) appear as members of four boards each, and five serve on two boards each. Of the total of 69 memberships on the 10 boards and commissions, 17 are held by four men and 27 by nine. The attempt to mitigate the multiplicity of agencies by repeating memberships is not to be overlooked.

A further consideration pointing in the same direction is found in the tendency for influence in the water management field to gravitate into the hands of a few individuals. Thus the executive director of one of the State's principal water agencies is a member of not fewer than half a dozen committees and subcommittees representing as many different agencies. One could compile a list of ten men, most of them at the top staff (executive director or executive secretary) level, in whom appear to center the principal initiative in basic state decisions in the water resources field. These are the men who, either personally or through assistants, attend the important meetings, prepare agenda for action and records of decisions taken, propose recommendations, and draft reports.

In view of the complexity of organization, one may be permitted to wonder whether the State's administrative machinery achieves all that might be expected of it in terms of program management. The Joint Legislative Committee on Natural Resources in 1953 delivered itself of this dictum:

> All of the above functions and agencies, blended together in one concerted program of action, have demonstrated New York State's approach to its natural resources and to their protection and use. This is a vast problem; the challenge to use these resources wisely and well is an equally vast one. It will continue to take the efforts of all administrative units of government and the interest and vigilance of the Legislature to assure the people of New York State that our state heritage of natural wealth will not be exploited for the present at the sacrifice of the future.[11]

One may embrace the Committee's conclusion without necessarily accepting its premise. The ultimate effectiveness of administration, assuming satisfactory program delineation beforehand, is not measured by official good will or by an internal sense of agency satisfaction but by the simple test whether the job to be done gets done well. It remains to be discovered whether "All of the [State's] functions and agencies [blend] together in one concerted program of action."

[11] *Report of the Joint Legislative Committee on Natural Resources* (Legislative Document, 1953, No. 69), p. 33.

CHAPTER IV

Water Resources Law

From the earliest days of recorded history, men have sought to regularize the use of water, and to evolve laws that would secure them in the enjoyment of their rights and protect them against misuse at the hands of their neighbors. The Code of Hammurabi, the oldest known set of written laws (about 2,300 B.C.), contained a provision to the effect that "If a man have opened his trenches for irrigation in such a careless way as to overflow his neighbor's field, he shall pay his neighbor in grain. . . ." Plato, almost 2,000 years later, disdained to undertake a codification of laws on the subject, noting that ". . . husbandmen have had of old excellent laws about waters, and there is no reason why we should propose to divert their course." Two excerpts will indicate something of the tenor of the existing laws summarized by Plato. "He who likes may draw water from the fountain-head of the common stream on to his own land, if he do not cut off the spring which clearly belongs to some other owner. . . ." ". . . If some one living on the higher ground recklessly lets off the water on his lower neighbour, and they cannot come to terms with one another, let him who will call in a warden . . . and . . . obtain a decision determining what each of them is to do." Some 800 years later still, the Institutes of Justinian (Book II, Title I) proclaimed that "All rivers and harbours are public; consequently the right of fishing in a harbour and in rivers is common to everyone." Further, "The use of river banks is public . . . , like the use of the river itself; and so every one is free to put in at the bank, . . . just as every one is free to navigate the stream. But the ownership of the banks . . . is vested in the riparian proprietors."

These laws, drawn at random from the legal systems of three separate civilizations covering almost 3,000 years, all treated of some of the common problems of water use. The principles stated have an oddly familiar ring, for they have counterparts in the law governing water rights which prevails in America today.

Consideration of New York's water resources law may well begin with mention of the Federal law on the subject. Such law includes relevant provisions of the Federal Constitution, congressional statutes and administrative regulations, and the major federal judicial decisions which have determined the breadth and depth of the written law in practice. The authoritative analysis of Federal water resources law lists the commerce power, the proprietary power, the war power, the treaty-making power, the general-welfare power, the interstate compact clause, and the clause extending the judicial power to ". . . controversies between two or more States . . ." as the principal constitutional bases for Federal involvement in water resource activities.[1] The examination extends its consideration of Federal interests at length under the headings of navigation, flood control, irrigation, power and multiple-purpose projects, other public purposes, related land uses, and comprehensive development. Chapter 10, book-length in itself, comprises a comparative summary. The volume contains a thorough analysis of the statutes passed pursuant to the granted powers, the administrative regulations promulgated in execution of statutory programs, and the principal court decisions interpretive of both. The mass of Federal law on water resources testifies to a long-standing and vigorous national interest in that field. When all is said and done, however, the national government remains one of delegated powers, even though a series of judicial decisions over the course of the last twenty-five years has minimized the significance of that concept for the field of water resources.

Of relevance also to an examination of New York's water resources law is the body of local law on the general subject of water which exists in the State. This law is embodied in state and municipal (and other local) ordinances, local administrative regulations, and the decisions of local courts not appealed. As a general rule, local lawmakers enjoy their powers by grant from the State, and state law is considered paramount. Policy therefore usually originates in state law, and local

---

[1] Report of the President's Water Resources Policy Commission, Vol. 3, *Water Resources Law* (Washington, 1950).

law plays the auxiliary part of facilitator. As such, its role is important, but it is not primary.

We turn then to the state as the repository of general governmental power from which traditionally flows the bulk of the substantive law governing water resources. Here six kinds (or sources) of law require mention. They are the constitution, the common law, statutes, administrative regulations, interstate compacts and agreements, and judicial decisions.

The Constitution

New York's Constitution contains the usual provisions for the protection of persons and property; and since water (more accurately, the use of water) is regarded as a property right, it is well to begin with those provisions. Two stand out among them. The first specifies that "No person shall be deprived of life, liberty or property without due process of law" (Article I, Section 6). The second stipulates that "Private property shall not be taken for public use without just compensation" (Article I, Section 7, [a]). Under the protection afforded by these provisions, the people of New York are secured against arbitrary interference by the State in their enjoyment of a wide variety of property rights in water.

More pointedly, the Constitution states categorically that "The use of property for the drainage of swamp or agricultural lands is declared to be a public use, . . ." and makes definite provision for the effectuation of this policy (Article I, Section 7 [d]). Article XIV treats of conservation, and Section 2 of that article provides that the Legislature may by general laws dedicate not more than 3 per cent of the Forest Preserve to reservoirs for municipal water supply and for state canals. As before, the section is explicit in stipulating how such reservoirs shall be constructed and operated. Article XV concerns canals; in three sections covering two full pages it sets out in some detail the practices to govern canal operation and use.

The provisions of the Constitution which, though devoted primarily to other subjects, relate directly or indirectly to water resources are numerous. Thus Article VIII deals at length with local finances, and Section 2-a concerns specifically local indebtedness for water supply and for sewage and drainage facilities. Another provision which bears directly on the subject of water is Subsection B of Section 5 of Article VIII, which excludes "Indebtedness heretofore or hereafter contracted to provide for the supply of water" in ascertaining the power of a local unit to incur debt. Section 5 of Article X treats of public corporations, and so concerns the problem of organizing for the administration of water programs. First and last, then, New York's Constitution gives considerable attention, both by direct provision and by implication, to water resources.

The Common Law

When the Dutch came to New Amsterdam in 1626, they brought with them the system of civil law. The original land grants therefore were governed by that law, as were the appurtenant water rights. Thus it came about that the rights in the waters of the Hudson and the Mohawk Rivers were governed by civil law principles, which were recognized as valid by the courts in a number of early cases.[2]

The coming of the English saw the common law superimposed upon the Dutch civil law, and in time the latter gave way to the common law. The Constitution clinches the transition by stipulating that such parts of the common law as are not repugnant to the Constitution shall continue to be the law of the State (Article I, Section 14). The courts therefore have a clear mandate to recognize and give application to the principles of the common law.

The courts, in defiance of the hydrologists,[3] have expounded a three-fold classification of waters. The first class includes natural watercourses and lakes. A natural watercourse is defined as a stream of water of natural origin flowing constantly or recurrently in a reasonably definite channel.[4] The second class consists of surface waters, which result from springs, rain, or melting snow, which lie on or flow along the surface of the earth, but which do not form part of a natural watercourse or lake. Percolating waters, comprising the third class, are sub-

[2] Joint Legislative Committee on Conservation of Water, *Report* (Legislative Document, 1912, No. 18), pp. 16-17.

[3] See Harold E. Thomas, "Hydrology vs. Water Allocation in the Eastern United States" (a paper prepared for the Symposium on the Law of Water Allocation in the Eastern United States, Washington, D. C., October 4-6, 1956), p. 8, where it is argued that ". . . (the) traditional legal classification . . . is scientifically unsound."

[4] The best manageable analysis of the common law of water rights and practice in New York is to be found in a report by William H. Farnham titled "Legality of Current Irrigation Practices." The report, prepared at the instance of the Joint Legislative Committee on Natural Resources, is to be found in the 1953 Report of that Committee (Legislative Document, 1953, No. 69), pp. 85-99. Heavy reliance is placed on the Farnham analysis in the summary which follows.

surface waters which percolate through the soil but follow no defined channel. Judicial classifications adopted elsewhere sometimes differentiate ground water in defined underground streams and lakes, which is associated with what the New York courts have denominated natural watercourses and lakes and which is distinguished from percolating water, which infiltrates the soil at random; and diffused surface waters, as distinguished from New York's natural watercourses and lakes, which are also surface waters.

The judicial interpretation of the law is clearest with regard to the second class; for the courts have ruled uniformly that the owner of the land on which surface water is found may use the water as he sees fit. They have held without qualification that the waters in marshes and ponds, and even in lakes surrounded by land held by one owner, may be used or consumed in any way the owner of the land on which they lie may wish. Since the right of the owner to water lying on his land appears to be absolute, there is no need to examine this branch of the law further.

Percolating waters (with which the New York courts associate what is frequently classified elsewhere as ground water) fall into a different category; for whereas in earlier years the courts accorded to the owner of the overlying land the same rights in percolating waters as he enjoyed in surface waters, the rule has come to be modified in recent years. The principal modification has occurred through an application of the rule of reasonable use which in effect means that, although the owner of overlying land may sink wells and pump therefrom any amount of water necessary for his own use, even though such pumping may interfere with the enjoyment of rights in the water by another owner, he may not withdraw excess quantities for sale off his land.[5] Inasmuch as the court specified that the owner has the right to take from his wells ". . . all the water that he needs [for] the fullest enjoyment and usefulness of his land as land," and further inasmuch as it listed as appropriate uses the ". . . purposes of pleasure, abode, productiveness of soil, trade, manufacture, [and] whatever else the land as land may serve," it is clear that the owner still enjoys broad rights to the use of "percolating waters" underlying his land.

A further limitation on the right to the use of percolating waters is found in the rule that a landowner may not, by pumping from wells located on his property, draw water from a natural watercourse or lake. If a well taps an underground natural watercourse or lake, or if

[5] The leading case on this subject is *Forbell* v. *City of N. Y.*, 164 N.Y. 522 (1900).

by percolation it draws water from a surface watercourse or lake, then the law governing water rights in natural watercourses and lakes applies.

Summarizing, the law with respect to percolating water permits the landowner to pump from wells on his property as much water as he may wish, subject only to the limitations that he use the water on his own land and that he not take water from a natural watercourse or lake in the guise of drawing percolating water.

With regard to natural watercourses and lakes the common law doctrine of riparian rights applies. In its pure form, the doctrine holds that the owner of riparian land has the right to have the stream flow past his land undiminished in quantity and unchanged in quality. Except for such uses as turning a water wheel which passes the water on unaffected in quantity and quality, strict application of this rule would render the water useless excepting to the ultimate downstream riparian landowner, that is, the last landowner before the stream enters the sea. Such a strict interpretation is applied nowhere in this country today. Instead a number of modifications have been introduced to make the doctrine workable in practice. In all states embracing the riparian system, it is worthwhile to note, it is not the water itself which constitutes the "property," but the right to use the water.[6]

In New York, the rights enjoyed by the riparian landowner are, in broad summary, three. First, he has a right to the continued flow of the stream past his land. "The right to have a natural watercourse continue its physical existence upon one's property is as much property as is the right to have the hills or forests remain in place."[7] Second, he has the right to make such use of the water as he can without injury to the equal rights of all other riparian owners, both above and below him. Third, he has the corollary right to the use of the stream's water without undue interference or injury by any other user. The essence of the riparian doctrine is the common ownership by all riparian landowners of equal rights in the use of the water of the stream or lake.

The riparian rights in all navigable waters are enjoyed subject to the over-riding limitation of the right of public navigation. Among

---

[6] Those who prefer a more splendid rendition may quote the report of a state agency to the effect that the right is ". . . an incorporeal hereditament appurtenant to . . ." the riparian land. Joint Legislative Committee on Conservation of Water, *loc. cit.*

[7] Henry Philip Farnham, *The Law of Waters and Water Rights* (Rochester, 1904; 3 Vols.), p. 1565. This monumental treatise continues to be a standard work after more than 50 years.

private rights, domestic use heads the list. A riparian owner may take from a stream sufficient water for all reasonable domestic uses, which have been interpreted to include household purposes, farm livestock, and gardens. Such uses are regarded as "natural," and they are paramount; for the domestic needs of all riparian owners have priority over all non-domestic uses.

After the domestic needs of all riparian owners have been satisfied, water may be utilized for non-domestic purposes. Among such uses is the production of power, either for industries or for the generation of electricity. Other important uses include water supply and recreation.

Non-domestic uses are subject to certain limitations, chief among which is the rule of reasonableness. Generally speaking, both purpose and amount of water used must be reasonable, the former in terms of accepted uses, the latter in terms of the right to substantially equal use by every other riparian owner. Here, indeed, is the crux of the riparian doctrine: that no use may be made of water which will substantially modify a like use by every other riparian landowner. The courts have held in a number of cases that what is reasonable depends on prevailing circumstances: volume of water available, gradient of the stream, the "usage of the country," the extent of the injury suffered by other owners, public necessity.

Application of the rule of reasonableness has led to certain generally recognized limitations. Among other prohibitions, the riparian proprietor may not:

1. take water from a stream for use on non-riparian land.

2. use water on land outside the watershed of a stream, even though the non-watershed land may be connected with the stream through riparian contact. (This is not a settled rule of New York law, but is the likely holding if the question should come up.)

3. take more than his reasonable share of water from a stream, even though he replaces his overdraft by water from another source.

4. impound water from a stream in a reservoir during a wet season with a view to using it during a dry season.[8] These limitations on riparian uses (there are, of course, others like them) add up to a restrictive interpretation of riparian rights in application of the rule of reasonable use.

Riparian law has been further modified through recognition of the doctrine of prescription. Under this doctrine, an adverse user (one whose use affects adversely the rights of other riparian owners) who

[8] Summarized from William H. Farnham, *op. cit.*, pp. 96-98.

persists in his use openly and continuously for fifteen years thereupon
acquires a perpetual right, which is judged to have become legal
through passage of time. An adverse, that is, an unlawful, user may
of course be interrupted at any time by appropriate legal action by
any aggrieved riparian owner.

Finally, it should be noted that riparian rights are enjoyed always
subject to the limitations of public action. Such action characteristically
takes one of two forms. First, the State may limit and in some instances
actually deny the enjoyment of private property rights in the exercise
of its police power, which may be defined as the residual power to take
action in behalf of the health, safety, morals, and the general welfare
of the people. In the case of a limitation on property use so severe as
to amount to appropriation, the injured party may have a legal claim
to compensation. Second, the State and its subdivisions, and certain
private agencies as well by specific grant, have the right to take private
property for public use on payment of just compensation. The police
power and the right of eminent domain are pervasive powers through
which the State may act to limit the use of or to appropriate private
property, including of course riparian property rights in the use of the
waters of natural watercourses and lakes.

Statutes

New York was concerned with water resources even in Colonial
times, and statutes on that subject were passed by the Colonial Legislature. The Constitution provides (Article I, Section 14) that such pre-
statehood laws as are not inconsistent with that document shall continue to be the law of the State, subject of course to modifications by
statutes subsequently passed. Since statehood, the Legislature has
turned its attention recurrently to the problems of water resources. To
cite a single area by way of example, there developed in the early
eighteen-hundreds a vigorous interest in "internal improvements," as a
result of which a mass of legislation dealing with canals was passed.
The domains of conservation and public health were favorite early
hunting grounds for the Legislature, which enacted laws on these subjects at almost every session. The statutes oftentimes were developmental or promotional in character—the canal acts were of this nature;
but frequently they were designed to regulate or limit the exercise of
private (and sometimes public) property rights in the public interest.
Almost all laws were individual acts passed without much reference

to what had gone before; but in the aggregate they bulked large in a number of programmatic fields by the late 1800's. Little attempt was made to codify or otherwise give systematic arrangement to statutory law until toward the end of the century.

With the passing of time the Legislature, and more particularly the responsible administrative agency, became conscious of shortcomings in the body of statutory law relating to a particular subject. Because of the piecemeal procedures employed, and more especially because of the *ad hoc* nature of the laws passed, inconsistencies inevitably appeared. Further, there were serious gaps in the law, since the sponsors of individual measures normally did not survey the statutory needs of the field as a whole. To cite a single example, between 1880 and 1892 the Legislature enacted into law some 50 chapters dealing with public health. The Board of Health (established in 1880) repeatedly urged the Legislature during those years to pass a comprehensive statute dealing with the whole field of public health; and in 1893 these efforts met success in the passage of the first general public health law.[9] The act repealed many fragmentary statutes and brought all public health legislation together in one inclusive law. It is a characteristic of a legal code that it is no sooner enacted than it is out of date, consequently a code requires periodic (or occasional) revision. Subsequent revisions of the public health law ensued in 1909, 1913, 1930, and 1953. Some of these revisions centered on administrative reorganization of the public health machinery, but all dealt to a greater or less extent with the substance of the law. The current Public Health Law (Chap. 879, Laws of 1953, as amended), excluding the Sanitary Code, fills (with annotations) a volume of more than 500 pages. The Sanitary Code, properly included as an appendix, adds another 150 pages. Logically arranged by articles and sections, the Law begins with a useful table of contents and concludes with a detailed index. The Public Health Law has many of the characteristics of a statutory code.

New York's effective statutory law is to be found in *McKinney's Consolidated Laws of New York, Annotated,* a ten-foot shelf of books running to 66 volumes. The great bulk of this law does not concern water resources, but a considerable percentage does. In 1953 William H. Farnham, then Dean of the Faculty of Cornell University, at the

[9] Chap. 661, Laws of 1893. Earl W. Murray's "Historical Development of the Public Health Law," which appears as a foreword to the *Public Health Law* (1954), affords a useful brief account of the growth of New York's public health law.

instance of the Joint Legislative Committee on Natural Resources brought together all statutes dealing with water resources.[10] This compilation (available only in typewritten form) found provisions directly or indirectly affecting water resources in almost twenty important fields of statutory law. It totals more than 600 closely-typed pages. Included are the relevant sections of such enactments as the Conservation Law, which incorporates the laws relating to the Water Power and Control Commission; the Public Health Law, containing the statutes covering the Water Pollution Control Board and its work; and the Public Authorities Law, which includes the acts pertaining to the Power Authority of the State of New York.

Administrative Regulations

Many state agencies have been given the statutory authority, in pursuance of their stated duties, to issue codes, rules, and regulations which have the force of law. Acting under constitutional and statutory directive, the Secretary of State has caused this body of law to be collated and published.[11] The resulting compilation runs to five large volumes, which contain regulations and codes adopted by no less than twenty-eight administrative agencies. In addition, there is a substantial annual supplement designed to keep the compilation up to date.

Among the best known of the administrative codes is that promulgated by the Public Health Council, which the law provides ". . . shall have power . . . to establish, and from time to time, amend and repeal sanitary regulations, to be known as the sanitary code of the state of New York, subject to approval by the commissioner" (*Public Health Law*, Section 225, Subdivision 3). The law further stipulates that "The provisions of the sanitary code shall have the force and effect of law and the violation of any provision thereof shall constitute a misdemeanor, punishable on conviction by a fine not exceeding fifty dollars or by imprisonment not exceeding six months, or both" (*Public Health Law*, Section 229). It goes without saying that any provision of the code is subject to challenge in the courts; many cases, indeed, have arisen to test the validity or the application of the Code's regulations.

Under the authority granted it by law the Public Health Council

[10] Joint Legislative Committee on Natural Resources, *Report* (Legislative Document, 1956, No. 63), p. 154.

[11] Secretary of State, *Official Compilation of the Codes, Rules and Regulations of the State of New York* (Albany, 1945, 5 vols.)

has developed a State Sanitary Code which comprises a detailed and rapidly growing body of regulations dealing with a great variety of subjects. Of special relevance to this study are Chapter V, which treats of drinking water supplies in a series of sixteen regulations; Chapter VI, which in twenty-three regulations deals with swimming pools and bathing beaches; and Chapter XI, which in Sections E and F sets forth the qualifications required of operators of public water treatment and purification plants and operators of public sewage treatment plants respectively.

Interstate Compacts and Agreements

New York is party to a number of interstate compacts and agreements, five of which are of direct significance for water resources. Of these, three are interstate compacts within the (Federal) constitutional meaning of the term. New York's participation in each instance has been signified by passage of an appropriate statute. The compacts, with the law of authorization in each instance, are the Interstate Sanitation Compact (Chaps. 3 and 4, Laws of 1936), the Ohio River Valley Water Sanitation Compact (Chaps. 776 and 945, Laws of 1939), and the New England Interstate Water Pollution Control Compact (Chap. 764, Laws of 1949). A fourth, the Interstate Commission on the Delaware River Basin, rests upon parallel statutes passed in each of the four participating states. New York's agreement to participate is incorporated in Chapter 600 of the Laws of 1939 and Chapter 610 of the Laws of 1949. The fifth, the Interstate Commission on the Lake Champlain Basin, until quite recently depended entirely upon voluntary cooperation; but in 1956 the Legislature of New York placed it on a firm (if temporary) footing by passing an act pledging participation by this State (Chap. 876, Laws of 1956). The interstate compacts enjoy a special status by virtue of their constitutional base, though in the end they depend on state acquiescence for their effectuation. The agreements have less status, being wholly dependent on acceptance by the participating states of decisions taken and recommendations made. In both cases the acts signifying state participation are basic to the success of the arrangement. The statutes cited here are symbolic of a new and developing field of law, and so are significant in New York's system of water resources law. The compacts and agreements themselves also are of basic importance, since they were adopted in each instance as a declaration of state policy regarding the subjects with which they deal.

Court Decisions

Last among the components of the State's water resources law may be noted the decisions of the courts. Such decisions are important for two major reasons. First, the common law depends for its elucidation upon the rulings of the courts in cases brought before them. Common principles exist, of course, in "the usage of the country," but their applicability to particular situations becomes clear only in the light of judicial interpretation. Court decisions declarative or interpretive of common law canons tend to be identified with and indeed to become indistinguishable from the common law itself. Second, the significance of a statute in practical application, and ultimately indeed its constitutionality, are proved by the courts as cases come before them for adjudication. The nuances of a given statute are never clear until the law has been tested by judicial action, frequently not once but many times. It follows therefore that court decisions must be reckoned as an integral part of New York's total body of water resources law.

AN APPRAISAL OF NEW YORK WATER LAW

An appraisal of New York's water resources law can be made at almost any desired breadth and depth. For present purposes, the discussion will be limited to comment on some of the principal features of the common law and of statute law, as the two relate to the water resources of New York.

The Common Law

The common law relevant to water resources concerns the doctrine of riparian rights. New York has embraced this doctrine in perhaps as pure form as any state, and the strengths and weaknesses of its basic water rights law therefore are largely those inherent in the riparian system.[12] The principal advantage claimed for the riparian doctrine in practice turns on its undoubted flexibility. This means, first, that adjustments can be made among conflicting uses in accordance with equity and need, and in harmony with the public interest. Second, the system rewards the enterprising and adventurous user, who may lose

---

[12] It should be noted that even in New York an element of appropriation practice creeps in through exercise of the authority of administrative agencies. The Water Power and Control Commission, as a single important example, allocates certain waters on the basis of proven need for specific purposes, subject of course to payment of compensation to riparian or other property owners.

his shirt but who also stands to gain from his daring. This is wholly in keeping with the American tradition of rugged individualism. Third, the flexibility of the system extends to the inauguration of new uses as well as to the expansion of old ones, and this permits further adjustment in the interests of both enterpriser and public. Fourth, precisely because no hard and fast set of rules exists, because there is no final spelling out of rights, the courts are enabled to take account of technological progress in their interpretation of the law, and to make the modifications required by changing conditions. It will be readily observed that all these advantages flow from the flexibility which is perhaps the prime characteristic of the common law.

Contemplation of the other side of the coin reveals that the riparian system also results in certain disadvantages in practice. Among these may be mentioned first the fact that the common law is administered by the courts, a feature worthy of elaboration. Judicial administration is negative by nature, since the courts on all except the rarest of occasions must bide their time until a case involving a point of law comes up in normal course. Then the action taken customarily is corrective only, and in any event it concerns only the question at issue. This means that judicial administration is also sporadic and piecemeal. It means further that it is also inconclusive, since the courts deal only with specific issues raised by the contestants and not with general problems or principles. The delineation of rights by the courts is a very time-consuming process, since years may be required to hear the individual cases necessary to explore the ramifications of a complex issue. This makes the judicial process a very expensive one for determining the extent (and the limits) of private rights in a given area.

Another major limitation on the courts as administrators is to be found in the extreme difficulty of effecting modifications in rules once established. Such modifications normally may be brought about only by the courts themselves in subsequent actions, and this is a long and tortuous process. Further, once the judicial tide has settled in a certain direction, it is difficult to effect a change in the prevailing position. Dean Farnham states boldly, with regard to one position long held by the courts, that "The legal reasoning on which the New York courts have founded this rule is fallacious, involving an erroneous conception of the relation to the problem of the doctrine of prescription." [13] It is of course not impossible for the courts to correct what Mr. Farnham regards as an erroneous doctrine; but seeing that the rule in question

---

[13] William H. Farnham, *op. cit.*, p. 97.

stems from a series of cases going back more than a hundred years, the prospect for modification does not appear bright.

Yet another shortcoming of judicial administration arises from the fact that judges are amateurs as administrators. Learning and experience in the law, however wide and deep, do not of necessity carry over into the vital field of administration. There both personal qualities and skills are necessary which are foreign to the requirements of the arena of judicial action. The argument is not that the courts do not do well what they are equipped to do, but that they do not and cannot be expected to do well what they are neither organized nor staffed to do.

On the substantive side, it is argued that the riparian system discriminates against the non-riparian landowner. Clearly it does this, for such discrimination is of the essence of the riparian doctrine. This may result in use of water where it is not needed and preclusion of use where it is needed, or where it could be utilized to better advantage. Another loophole results (or may result) in waste, in that water not claimed by riparians (particularly lower riparians) may be allowed to flow unused to the sea.

But perhaps the most serious charge brought against the riparian system grows from its most vaunted advantage; for the medal which reads "flexibility" on one side, reads "uncertainty" on the other. This uncertainty extends to the farthest capillaries of water rights practice, for a user can never be sure that a right enjoyed by him, however long, is in truth a legally defensible right in all circumstances. He may be certain, on the contrary, that it is not a legally defensible right in all circumstances, since any right he enjoys is subject to the assertion of a like right by every other riparian owner and so subject to intrusion through judicial modification of earlier rules.

If the established riparian user must pay for the flexibility of the common law by suffering a precisely equal measure of uncertainty, the would-be new user is in an even greater quandary. He confronts a use system long ensconced, protected both by tradition and by the courts, and geared for defensive action. That this picture is not overdrawn is witnessed by the unenviable position occupied by irrigation as a potential new user in New York today. As late as 1950, irrigation for agricultural purposes required negligible amounts of water. Since that time the use of water for irrigation has increased rapidly, and the potential requirements are great.[14] New York's riparian system is ill

[14] Temporary State Commission on Irrigation, *Report* (Legislative Document, 1957, No. 27), pp. 72-75.

adapted to the accommodation of such an important new use; for while the common law has always recognized as primary such limited farm uses as stock-and-garden-watering, it has developed almost no rules with regard to irrigation for the reason that that particular use has assumed importance in the State only in the last half-dozen years. The riparian system as it operates in New York appears to be almost wholly incapable of meeting the new demands of irrigation in any reasonable time. Irrigation facilities are expensive, and farmers need to know what they can count on before making the heavy investments required for permanent storage, pumps, and conduits. Since under the present system they do not know, and since they are not prepared to assume the risk arising from the uncertainty, it is fair to say that the law serves as a serious deterrent to the development of irrigation. The Temporary State Commission on Irrigation concluded early in its deliberations that "a clearer definition of water rights (with respect to irrigation) seems imperative." [15]

Where does the advantage lie, under the riparian system, in the continuing contest between flexibility and uncertainty? This is the main issue to be resolved in weighing the advantages and disadvantages of the common law doctrine of riparian rights.

Statute Law

One close to the scene recently stated that several hundred bills on conservation are introduced into the Legislature each year, of which 50 to 75 usually are enacted into law.[16] Like numbers are introduced concerning other important substantive fields, and some hundreds of measures find their way into the statute books at each legislative session. What proportion of these enactments relate to water resources no effort has been made to determine, but the gross water-law product disclosed by William H. Farnham's compilation (mentioned above) indicates that it is substantial.[17]

[15] *Ibid.*, p. 75.

[16] J. Victor Skiff, Deputy Commissioner of the Conservation Department, in his foreword to the volume on *Conservation Law* (Book 10 of *McKinney's Consolidated Laws of New York, Annotated*).

[17] Evidence on the water-law product of one state is found in the testimony of Governor Goodwin Knight of California, who in March of 1957 estimated that, of the 7,000 bills, constitutional amendments, and resolutions then pending before the Legislature of that State, approximately 10 per cent had a direct bearing on water resources. A prominent New York administrator says this figure is entirely too high for New York. From 1 to 2 per cent, he feels, would be nearer right.

Whatever the bulk of this mass of water legislation, its chief feature is that it is largely unplanned and unguided. It is not fair so to characterize all water measures, for there are bodies which exert an influence toward the integration of legislative proposals over certain specific areas. It remains true nevertheless that most water bills are introduced hit-or-miss, in an effort to meet a felt need or relieve a demonstrated pressure.

One result of this indiscriminate system of legislation (which is by no means peculiar either to the water resources field or to New York) is that, since there is no agency or body responsible for surveying the whole field of water law, some areas of that subject have received quite inadequate attention—some segments are seriously under-legislated. The Legislature passed a very important act dealing with water power in 1943. In his memorandum of approval, the Governor observed that the need for such legislation had existed for many decades. Noting the act at hand as a "major step forward," he expressed the hope nevertheless that there might soon be ". . . a complete revision of legislation affecting water power . . . to make the system of state control more efficient, more economic and to avoid the possibility of unjust private exploitation of the resources that belong to all of the people." [18] Water power, it should be noted, was a field in which, in 1943, the State had been actively interested for almost four decades. Another important subject which is knocking at the door of the Legislature at this moment is irrigation, concerning which on the one hand the State's system of riparian law is largely preclusive while on the other there has been no statutory redress.

Note has been made of the obsolescent nature of a statutory code, and of the need for occasional revision if a compilation of laws is to be kept in useful form. A number of New York's compilations containing important water resources laws (there is no single compilation or "code" in that field as such) are in want of a thorough revision. The nature of the problem is indicated by the concurrent resolution establishing the Joint Legislative Committee on Revision of the Conservation Law (1955). The resolution begins:[19]

WHEREAS, There has been no general revision of the conservation law since the general revision of nineteen hundred twenty-eight; and

WHEREAS, Many amendments and changes have been made over the years to such law resulting in inevitable ambiguities, weaknesses and defects; and

[18] *Conservation Law*, pp. 456-457.
[19] Assembly Resolution No. 25, 1955.

WHEREAS, It is desirable and needful that such law be clarified, re-codified and generally revised; now, therefore, be it . . .

Twenty-seven years is a long time to wait for a revision of the general law in so important and active a field as conservation—and, of course, the authorized revision is still far from completed.

New York's haphazard legal approach to the problems of water resources is reflected in the fact that the State thus far has not seen fit to make a general declaration of public policy with respect to the subject. There are at least two policy declarations of limited scope. The first asserts the permanent right of the State to exercise power and control over the waters in which it has the proprietary ownership of the flow and to the use of which it has paramount and exclusive right (Chap. 46, Laws of 1943). The second declares it to be state policy ". . . to maintain reasonable standards of purity of the waters of the state consistent with public health and public enjoyment thereof, the propagation and protection of fish and wild life, . . . and the industrial development of the state, and to that end require the use of all known available and reasonable methods to prevent and control the pollution of the waters of the state of New York" (Chap. 666, Laws of 1949). There may be other similarly limited statements of legislative intent (though these are generally recognized as the two most important), but there is nothing approaching a comprehensive declaration of public policy regarding water resources.

In view of this deficiency, one is not surprised to discover that there is no separate compilation of statutory water law. There are many water laws, but they are scattered through several volumes of the Consolidated Laws; they are found in such aggregations as the Conservation Law, Public Health Law, Navigation Law, Canal Law, and Public Authorities Law. The unofficial compilation made for the Joint Legislative Committee on Natural Resources (mentioned earlier) is evidence of an active legislative interest in water resources, an interest which, however, has yet to be coordinated and rationalized.

The state of New York's water resources law, common and statutory, reflects several important facts. First, the State has bountiful water resources, and (with the notable exception of New York City) need has only recently begun to press against supply—and then only in limited areas. Second, although New York has many excellent individual statutes on particular aspects of the subject, it is just beginning to give significant attention to the problem of water resources as such. Third, and as an almost inevitable corollary, New York's system of water resources law is in an amorphous state. Partial and uneven development has sufficed in the past because, with certain well-defined exceptions, the problems have not been overly complex nor the demands for concerted attention imperative. Within the last ten years the use:supply relationship has changed to produce rapidly increasing pressures on resources. The day for a legal reckoning may be at hand.

THE MOVEMENT FOR LEGAL REFORM

Water resources law is in a state of change throughout the country. The eastern system comes under more and more heavy attack as water needs press on supplies; while the western system, designed from the beginning to deal with a short-supply situation, is criticized on the ground that it has failed satisfactorily to achieve its purpose. It is perhaps not unnatural that each system should seek to draw strength from the other, nor that the two therefore should seem to be approaching a common ground.

There are two basic systems of law for the determination and protection of water rights in the United States. The first of these is the riparian system, whose principal features have been described and evaluated with reference to New York practice. The riparian doctrine prevails throughout the east, and is found in all the thirty-one states east of the Mississippi River with one quite recent exception. The riparian system has been modified in details from state to state, but in essential characteristics it remains substantially uniform in content and application. Its advantages, as earlier noted, are principally those which stem from its flexibility; its major disadvantage is the companion characteristic of uncertainty, together with the limitations inherent in a system of administration by the courts.

The seventeen states west of the Mississippi are said to employ the system of prior appropriation of water rights. Actually, only the eight intermountain states (Arizona, Colorado, Idaho, Montana, Nevada, New Mexico, Utah, and Wyoming) employ prior appropriation to the exclusion of riparian rights. The remaining nine western states (those of the Pacific Coast and the Great Plains) employ a blend which recognizes certain basic features of both systems. The distinction between the prior appropriation system in its pure form and the combination adopted in these nine states, while not unimportant, is quite technical. Further, the appropriation system appears to be in the ascendancy throughout the west. The present discussion is limited to consideration of the prior appropriation system.

This system originated in the west of the mid-nineteenth century, where water was scarce and laws were rudimentary or oftentimes non-existent. It grew from the practice of the early settlers of taking water where they found it and using it where they needed it. In the course of time, and by common agreement, a rude system of notice was adopted to indicate the amount taken by and the nature of the use of each appropriator. The features of the prior appropriation system, as they developed through pioneer practice and as they prevail, with refinements and some variations to this day, are these:

1. Water rights are regarded as a function not of place but of use. They therefore are not limited to any particular class of landowners, riparian or otherwise, but are available to any who can show need.

2. Need is defined in terms of beneficial use. Water not only must be put to beneficial use at the time of appropriation; it must continue to be used beneficially, else it is subject to reappropriation. (Aside from domestic use, mining and irrigation were the principal uses to which water was put in the early days. At present 90 per cent of all water used in the west is consumed in irrigation.)

3. "First in time, first in right." Application of this principle has led to gradations in appropriators, since clearly "senior" appropriators would have priority of rights over "junior" appropriators in the event of water shortage through over-appropriation or through drought. The concept accounts for the word "prior" in the term "prior appropriation system."

4. Self-governance in respect of appropriation rights early gave way to public governance. The trend is toward administration by a state agency, although a few states still rely primarily on the courts for determining water rights. In those states which have established administrative units, the usual functions of the agency include receiving and passing on applications for new uses, inspecting to see that laws and regulations are being observed, (in some states) administering the water distribution system, and participating in the adjudication of conflicts.

In summary, the principal differences between the riparian system and the prior appropriation system are these: (1) the one emphasizes *place* of use, the other *beneficial use* without reference to place; 2) the one allocates water rights equally among all riparian owners, allowing the assertion of a new or previously unused riparian right at any time, the other appropriates the available water on a priority basis to claimants for beneficial uses in fixed amounts and for all time; and (3) the one relies on the courts to administer water rights, the other vests (or tends to vest) the functions of water management in a special administrative agency.[20]

The tremendous increase in the use of water in the last few years has posed new problems even for the states which had previously considered their water resources adequate to all foreseeable needs. The growing concern over water resources is reflected in a resolution adopted by the 1954 General Assembly of the States which reads thus:

> The various problems relating to the use and conservation of water resources and rights thereto need review and possibly revision in the light of increasing demands upon water resources. The twelfth General Assembly of the States requests that the Council of State Governments and its Drafting Committee of State officials consider recommendations, including model legislation, as to the improvement of state water law. This consideration shall include, among other things, appraisal of the merits and applicability of the various legal doctrines relating to water rights and the possible evolution of new legal principles in this field and development of legislation fostering watershed organization for management purposes.[21]

The Council of State Governments, in harmony with the spirit reflected by this resolution, has made water resources law one of its major concerns since 1954.

The states have reacted variously in the face of the impending emergency. In the west, where the water shortage is perennial, they have sought to improve not only the law but also the administration of water resources. In the east, where the supply of water has always been assumed to be adequate to almost any need, a number of states have made studies of their water resources problems and have found cause for concern. Some states, Indiana, Kentucky, and Virginia among them, have sought to clarify and strengthen the riparian system by statutory enactment. Several others have considered legislation designed to impose the appropriation system on the existing system of riparian rights; and one, Mississippi, passed an act in 1956 which made

[20] Henry Philip Farnham, *op. cit.*, contains a very useful discussion of the origin and practice of the system of appropriation. See Vol. III, Chap. XXII. Recent analyses of the prior appropriation doctrine are to be found in two papers prepared for the symposium on the law of water allocation (Washington, October 4-6, 1956). See Clyde O. Fisher, Jr., "Western Experience and Eastern Appropriation Proposals"; and Charles M. Haar and Barbara Gordon, "Legislative Change of Water Law in Massachusetts: A Case Study of the Consequences of Introducing a Prior Appropriation System."

[21] *State Government*, Vol. XXVIII, No. 1 (January, 1955), p. 27.

it the first state east of the river to adopt the appropriation system.[22]

The increasing interest of the east in the appropriation system is unmistakable. The reason for this phenomenon lies, of course, in certain presumed advantages which that system enjoys over the traditional system of riparian rights. Among these, the principle of full utilization of the water resources deserves to be ranked first. Second, the appropriation system permits the use of water where needed without tying use to riparian ownership. Third, it restricts the appropriator to beneficial use, both as to purpose and as to amount, and makes possible the reappropriation of excess amounts not beneficially used. Fourth, the amount, purpose, and place of use of appropriated water are definitely indicated, and this adds an element of certainty which is unknown under the riparian system. Fifth, management of water resources by the courts through individual case rulings gives way under appropriation to management by a full-time administrative agency staffed by experts.

On the other side of the ledger, the appropriation system has certain disadvantages, most of them arising from problems not solved or only partially solved. The first of these is a concomitant of the certainty that is so greatly desired, for a system which is quite likely to be somewhat inflexible. In this instance, the inflexibility arises from the feezing of rights in terms of the original appropriation. This may lead to water use which is not economically or socially sound. Thus poor land may have been devoted to farming early and so may have pre-empted water rights ahead of potentially more productive land which arrived at the stage of cultivation too late to qualify. Moreover, the prior appropriation of so much of the available water for irrigation militates against the re-allocation of water to urban/industrial uses. *Priority* and *beneficial use* are the twin hallmarks of the appropriation system; but as between the two, priority tends to become the more important in practice.

This heavy emphasis on priority results in a second major disadvantage, one associated with the problem of waste. The conditions under which an appropriation was made often change, so that the water appropriated is no longer needed in the amount, or at the place, or for the purpose specified, yet the appropriator continues in full enjoyment of his right of use. This is not the way the law has it, but it is the way the system frequently works in practice.

A third problem arises from the difficulty of harmonizing the ri-

[22] *Advance Sheet General Acts, No. 10* (Mississippi), Regular Legislative Session, 1956, House Bill No. 232.

parian and the appropriation systems. This problem has not arisen in the inter-mountain states, where the riparian system is not recognized; but it has proved to be quite stubborn in the nine states which employ a blend of the two systems. It would pose many issues for an eastern state adopting the appropriation system. Mississippi, in its statute embracing the appropriation system, has provided for a combination of the two; but while this is an entirely logical resolution of the problem in principle, it gives rise to many complex issues in practice, as California's long experience will attest.

A fourth major difficulty arises from the problems inherent in the administration of the prior appropriation system. The management of a state's water resources requires a devotion to duty, a level of technical competence, and an imperviousness to pressures not always found in satisfactory combination. This point is the more significant in view of two considerations. First, several of the states employing prior appropriation do not have general statutory career services (merit systems), and their employees are less experienced and less competent than could be desired. Second, notwithstanding the doctrine of state administration, actual management frequently falls to local water masters named either by county governing bodies or by irrigation district boards, the state official nominally responsible merely ratifying the selections. This would not seem a particularly reliable process for bringing expertise to bear on a state program. This line of argument nevertheless largely begs the question, for many activities of government, public health and education among them, make equally severe demands upon their administrators. A plan or method is hardly to be rejected because it requires integrity, courage, and competence of its administrators.[23]

On balance, what appear to be the relative advantages and disadvantages of the riparian and the prior appropriation systems, from the point of view of the eastern states in which the comparative merits of the two are now under consideration? While it would be presumptuous to propose a categorical answer to this question, certain relevant considerations can be pointed out. On the one hand the flexibility *plus* the uncertainty of the riparian system must be balanced off against the certainty *plus* the tendency toward sclerosis which characterizes the appropriation system. On the whole, the advantage here may well lie with the appropriation system, which in the coming days of shortage

[23] Clyde O. Fisher, Jr. (*op. cit.*) has made a careful appraisal of the relative advantages of the two systems, and more especially of the points of strength and weakness of the appropriation system.

would provide both private users and public the assurance which will be needed regarding the amount of water available to a particular user and the place and purpose of use.

On the other hand, the decision may turn on administrative considerations. Those who place their reliance in the adversary system will take their stand with the common law and its time-honored process of case-by-case administration by the courts. By the same token, those who question the competence of the courts for administration, who regard the management of water resources as an increasingly technical problem requiring the continuing attention of professional administrators, who, in short, favor the administrative over the judicial approach to the problem of water resources management, are likely to prefer the appropriation system.

Charles M. Haar and Barbara Gordon have argued that management by administrative professionals is not necessarily peculiar to or inseparable from the appropriation system.[24] In their view, the major advantages of an administrative system might be incorporated into riparian practice with less disjunction than if the states (Massachusetts, in this instance) were to adopt the appropriation system outright. This might be done, they argue, through the assumption of a more active role by the courts, particularly in the direction of making more general use of the special master system. "If there is not already specialization among masters in a state there seems to be no reason why there could not be in the future. This permits the judiciary to become expert by the process of continuous intervention. Nor is there any reason why the master handling water cases could not have a permanent staff of hydrologists, economists, and physical planners, thereby tapping the latest scientific knowledge and techniques concerning water uses." [25] The advantages of a system of administrative supervision thus are recognized by at least some who defend the common law system of riparian rights; and this may well be the residual deposit of the argument.

New York has not been among the eastern states expressing active interest in a fundamental revision of its water rights law. It is true that there has been some dissatisfaction over the doctrine of riparian rights as it has developed here through judicial application. The Farnham report (1953) was generally critical in tone.[26] The Joint Legislative Committee on Natural Resources, through its sponsorship of the Farn-

---

[24] Charles M. Haar and Barbara Gordon, *op. cit.*
[25] *Ibid.*, p. 34.
[26] William H. Farnham, *op. cit.*

ham report, offered evidence of its interest in water rights law, and in its 1955 recommendations to the Legislature gave tangible statement to that interest in these words:

> There is need for a thorough evaluation of the water resources available in New York State to serve the needs of municipalities, industries, agriculture and private water users. Concomitantly, there is need for a clarification of the rights of all persons to the use of the waters of the state. Every effort must be made to recognize and protect the rights of those who benefit from the use of waters of adequate quantity and useful quality.
>
> It is recommended that studies of this problem be initiated by creating a special sub-committee of the Joint Legislative Committee on Natural Resources, composed of representatives of various interested agencies and groups to consider proposals for a basic water rights policy in New York State.[27]

The genuineness of the Committee's concern is not to be doubted, but its activities have been limited to studies, resolutions, and reports. Neither the Committee nor any other official body has found need thus far for a fundamental change in the State's system of water rights law.

This side of proposals for fundamental change there is considerable activity. The Joint Legislative Committee on Natural Resources, the Joint Legislative Committee on Interstate Cooperation, the Joint Legislative Committee on Revision of the Conservation Law, the Temporary State Flood Control Commission, the Temporary State Commission on Irrigation, the various interstate (water) commissions of which New York is a member—all these and very likely others as well are engaged in the business of drafting water resources legislation. Among these it may be said, without intention to minimize the energetic efforts of some of the others, that the Joint Legislative Committee on Natural Resources is the most active. Following its 1955 resolution (mentioned above), the Chairman of that Committee appointed a Special Advisory Committee on Water Resources and Water Rights.[28] It is too early to tell what direction the work of this advisory committee will take or what fruit it will bear. More nearly than any other body, this committee represents New York's continuing concern for the

[27] Joint Legislative Committee on Natural Resources, *Report, 1955* (Legislative Document, 1955, No. 76), p. 229.

[28] Joint Legislative Committee on Natural Resources, *Report, 1956* (Legislative Document, 1956, No. 63), pp. 155-156.

general subject of water resources law. Its work will be watched, therefore, with special interest.[29]

Meanwhile, it is well to remember that New York's interest in its water resources as such is relatively new, that its concern finds expression through many channels, and that its administrative structure in the field is seriously splintered. It may be questioned whether water resources as a problem requiring separate and special attention has not come of age, and whether, if the conclusion is affirmative, the issue of water resources law is not appropriately to be regarded as a part of the larger problem. In the concluding chapter of this study the recommendation is made that a state water resources commission be appointed to survey all aspects of the water problem and bring forth an integrated report on the subject. The law of waters and water rights should be recognized by the commission as a facet of the total problem worthy of its special consideration.

[29] It is worth noting that up to this time a major portion of the efforts of the Joint Legislative Committee on Natural Resources has been devoted to the problems of the New York State Forest Preserve, which was established in part for the protection of the water resources. In these activities also the Committee's work bears upon water-resource administration, though by definition watershed protection and management are beyond the scope of the present study.

Part II: Action

Introduction

To THIS point, the discussion has confined itself to what might be called the anatomy of water resources administration in New York: the water programs which, as a matter of public policy, New York has determined to carry on; the various administrative organizations through which programmatic activities are pursued; and the system of law on which water rights and water practices rest. This has proved useful, for it has provided a background for the further development of the subject. It has set a stage, but so far the stage is without actors.

It is desirable now to add a second dimension, that of action, to the analysis. A description of framework serves a purpose, but it is not fully useful until considerations of time and motion are added. To the concept of anatomy the scientist desirous of understanding an organism adds the further concept of physiology, which concerns the living tissue, with its circulatory system, surrounding the skeletal structure. Applying the figure to the problem at hand, it is necessary to proceed now to an examination of the methods of work of the administrative organization, in terms of the several major public programs. How do the many agencies operate? What are their relations with one another? With Federal agencies in the field? With local organs? With clientele groups? The overriding and ever-present question of course concerns programmatic responsibilities and the methods of their discharge. If we can arrive at comprehensive answers to these questions, we shall have acquired a basis for an understanding of New York's manifold activities in the water resource field.

There are several ways of approaching such an operational analysis. One approach (the one most commonly employed) emphasizes a point-by-point examination of the tasks assigned to each agency. This may be called the taxonomic method. Relying on such sources as statutes and public reports, the researcher proceeds systematically from agency to agency, classifying and recording the duties discovered, until at length all the agencies are exhausted. A major shortcoming of this method is that, since it treats of movement only at a distance and, so to speak, second-hand, it leaves an impression of administration as a static thing. Another is that the reader usually is exhausted before the agencies are.

A second major method of analysis utilizes what has been called the case approach. Here the focus is on program, and more especially on the development and resolution of a particular programmatic situation. The analysis develops in narrative fashion as the problem takes shape; administrative agencies move in and out as their responsibilities or interests wax and wane; issues assume form and meaning in the struggle which ensues; and resolutions emerge as the sound and fury dissolves in compromise. Administration is essentially the activity of officials in motion. Case analyses get (or should get) behind the paper curtain of formal administration into the realities of human relationships. The case method of developing and treating data, many believe, gets closer to the core of administrative action than any other. It is the method to be employed here in an effort to add the dimension of physiology to the anatomical structure described in Part I.

The first problem to arise in connection with case analysis concerns the materials to be selected for examination. The incidents which might be chosen to illuminate administration are literally without number: a recent writer on the subject used the phrase "an embarrassment of riches." Every decision reached, every hearing conducted, every petition handled, every contract let, every order entered—in short every measure taken by an administrative official contributes to that aspect of government called, in the aggregate, administration. It follows that knowledge about any and all of these measures contributes to understanding of the administrative process, that is to say of government in action. But it would be manifestly impossible to examine the whole realm of administrative action within any reasonable limits of energy and time, not to mention interest and patience; hence it is necessary to select examples for analysis. This must be done with discrimination, for both substantive action and agency (or agencies) involved must be in the mainstream of administration.

Recognizing that, since no two sets of circumstances surrounding an incident or occurrence will ever be the same, there is no such thing as a "typical" case in administration, four more or less representative administrative situations nevertheless have been chosen for exploration in

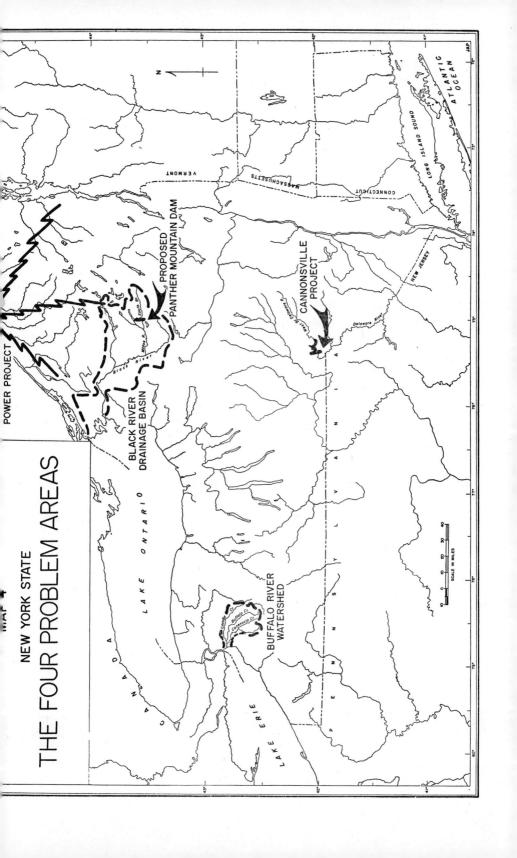

MAP 4

NEW YORK STATE

THE FOUR PROBLEM AREAS

POWER PROJECT

PROPOSED PANTHER MOUNTAIN DAM

BLACK RIVER DRAINAGE BASIN

CANNONSVILLE PROJECT

BUFFALO RIVER WATERSHED

SCALE IN MILES

LAKE ERIE

LAKE ONTARIO

CANADA

PENNSYLVANIA

NEW JERSEY

CONNECTICUT

MASSACHUSETTS

VERMONT

LONG ISLAND SOUND

ATLANTIC OCEAN

depth. Each concerns a major aspect of New York's water-resource programs, though as many as three relate secondarily to other programmatic activities as well. Among them they involve a number of governmental units and divisions, including the State's principal water-resource agencies. The first turns on the problem of water supply; it has to do with the intricate procedures surrounding New York City's successful effort to bring water to the city from the headwaters of the Delaware River high in the Catskill Mountains. The second involves a furious controversy which developed over an attempt to bring under control the Black River in the north central part of the State. The third finds its locale in the west, where a program of pollution abatement for the Buffalo River has just been launched. The fourth, centered on the St. Lawrence Seaway development in the north, concerns the allocation of hydroelectric power from the Barnhart Island plant. Map 4 indicates the location of the four problem areas.

The central issues to be examined are, therefore, water supply, river regulation, stream pollution, and hydroelectric power allocation. The numerous organs involved represent every level of government, though by design the spotlight remains on the state agencies. Cliental groups in considerable number appear as special pleaders. The cases portray government in action in a variety of administrative situations. To advert to an earlier metaphor, the cases of Part II put actors on the stage set by Part I, provide them with lines to speak and parts to play, and establish plots whose unfoldings are designed to reveal what New York's major water-resources agencies do and how they do it.

CHAPTER V

Water Supply: The Cannonsville Reservoir

As recently as 1890, only one-third of the people of the United States lived in urban places; today the proportion is two-thirds. One of the most striking phenomena of our national growth has been the rise of the cities, and more particularly, during the last two decades, the rise of the great cities. From 1950 to 1956, 85 per cent of the total population increase of almost 15 millions occurred in and around the larger cities. The Twentieth Century Fund recently characterized the whole eastern seaboard from Boston to Norfolk as an emergent metropolis, and bestowed upon it the name of Megalopolis. Among the many grave problems attendant upon this unexampled development is that of water supply; for it is the fate of *homo metropolensis* to wander the asphalt desert, cup in hand, in search of water to slake his thirst. This is true particularly of the denizens of New York City, which has spent a good share of its time and treasure since Manhattan was New Amsterdam in quest of a suitable water supply.

WATER FOR THE CITY

New York's Early Search for Water

In the beginning of settlement on Manhattan Island, the citizens were left to shift for themselves in the matter of water supply. This they did through private wells and pumps. As early as 1658, the Dutch dug a public well in lower Manhattan, and when the English took over they followed suit with another half-dozen wells for the common use. During the next hundred years the City's water supply system underwent changes in two directions: first, the number of wells, both public and private, increased enormously; and second, the quality of water deteriorated to the point where, it was reported in 1748, even the horses refused to drink it.[1]

Early experience with the provision of water through collective effort proved either futile or almost wholly unsatisfactory. Philadelphia concluded early to entrust its water supply problem to public hands, but New York refused to follow that lead, placing its faith instead in private enterprise. The first important effort to establish a general supply was brought to an untimely end by the Revolutionary War. The Legislature of New York launched the second by granting a charter to the Manhattan Company in 1799. That company, though chartered for the purpose of securing to the City an adequate supply of good water, turned out in actuality to be a device by which Aaron Burr and his Republican collaborators gained entry into the banking business, theretofore held as a close monopoly by the Federalists. The company did furnish some water of a sort, by way of complying with the terms of its charter, but its service was never satisfactory and it engaged almost continuously in a running battle with the City for more than a third of a century. Meanwhile, a constant procession of proposals filed by—an abortive scheme to bring water into the City by canal, a bored-well enthusiast who generated considerable support, a spring water company that failed to make the grade, a general water-works corporation which was forced by its stockholders to dissolve before it could even test its shaky legal legs. Meanwhile there followed, too, a succession of epidemics and fires which emphasized recurrently the need for an adequate water supply.

The State had manifested an interest in the City's water problems from an early date. Note has been made of the legislative charter to the Manhattan Company in 1799. Thereafter almost every session saw some measure introduced relating to water for the City—a petition for a charter for another corporation, a memorial to rescind a charter already granted, a proposal to authorize the issue of new bonds for exploration or development, a demand that such-and-such enterprise be investigated. More than once the Legislature acted favorably on a proposal, though never (until the 1830's) in a way that promised any real relief.

Meanwhile, tension in the City built up to the point of explosion. Unusually heavy fire losses in 1828 were followed in 1832 by an epi-

---

[1] Nelson M. Blake, in *Water for the Cities* (Syracuse: Syracuse University Press, 1956), has written a useful and highly readable history of the search of the early cities for water. This section draws heavily on his account, not only for substance but, as will have been observed, for title as well.

demic of cholera which claimed 3,500 lives. The Manhattan Company had long since demonstrated its incompetence (or its unwillingness) to deal with the problem and relief more and more came to be sought from other quarters. The recent years had yielded a number of engineering (and other) surveys, most of which recommended that water be brought in from the outside—some preposterous proposals would have had the City go as far away as the Croton River for its water. Meanwhile the city leaders, despairing of satisfaction at the hands of private developers, turned to the Philadelphia example and began to think and talk seriously of public action.

These several developments came to a head in the decision of the City's Common Council to request legislative authorization to proceed in behalf of the City. This request was granted when, in 1833, the Legislature passed an act providing for appointment by the Governor of a board of five water commissioners. A second act, passed in 1834, authorized the City to issue water stock to the amount of $2,500,000 to defray the cost of a water system, and vested authority to plan and construct a system in the water commissioners. By this time opinion, engineering and otherwise, had pretty well settled on the Croton River source and the commissioners soon determined to proceed with the Croton dam-and-aqueduct scheme. In a vote taken in 1835 the citizens of New York approved the proposal, and the last procedural barrier was removed. The last lingering doubts disappeared in the smoke of the City's worst fire later the same year. The need for water was imperative, and the City was at last ready to go after an adequate supply.

The years 1836-1842 were spent in constructing the Croton dam and aqueduct, but that is a very prosaic way of telling what happened. First the owners of the properties involved, not to mention a good number of citizens not property owners, protested vehemently against the invasion of their rural countryside. Their qualms were not lessened by the army of immigrant workers who overran the damsite and the aqueduct right-of-way all the way down to the City. The water commissioners found it necessary to replace their chief engineer early in the game, only to be themselves replaced somewhat later with a change of administrations in Albany. The resulting wrangle between the Whig water commissioners and the Democratic city government yielded no advantage to the project. The original cost estimates proved much too low, which caused the Whigs to exult and the Democrats to scrounge for new funds. The commissioners, whose engineering staff had recommended a low bridge to convey the water across the Harlem River, were directed by legislative act to build a high bridge instead. The new dam proved unequal to the strain placed upon it by an unprecedented flood and washed out, with some loss of life and considerable loss in property. Fortunately there was time to rebuild the dam while the aqueduct was being completed.

The great day arrived when, in the spring of 1842, the water commissioners announced that both dam and aqueduct were ready. The City celebrated the arrival of Croton water in its reservoir on June 27; it combined the traditional July 4 observance with a further celebration of its new water supply; and finally it topped off these preliminaries with a grand water rally on October 14. A splendid parade was followed by addresses by all and sundry, including principally Governor Seward and Mayor Morris. The board of water commissioners had its day at last, and it was a long day, beginning with the sunrise and lasting until nightfall. The celebration was indeed worthy of the event. New York City for the first time in its history had an adequate supply of good water. The City at that time had a population of well above 300,000.[2]

Those who had been responsible for pushing the Croton aqueduct through to completion had reason to believe that they had planned well. All supposed that the new system would prove adequate for many years to come. The Whig water commissioners, who had inherited what they considered to be a grandiose scheme from their Democratic predecessors, were of the opinion that the reservoirs which the plans called for would not need to be fully utilized for a century, and maybe not ever. Yet within twenty years water was once more in short supply. The need was met temporarily by enlarging the aqueduct at certain points where economy had been practiced in construction, building a huge new reservoir in Central Park, and, a little later, erecting a new and much larger Croton dam. Further, modest new sources were acquired by tapping two small streams east of the Hudson. These measures relieved immediate pressures but afforded no long-range solution. It was clear by the end of the century that important new supplies must be found. The extension of the City's limit to include the surrounding boroughs both exacerbated the problem and confirmed the conclusion.

As so often in the past a number of questionable proposals were made for dealing with the problem, and as so often before the City somehow threaded its way through these booby traps to a workable

[2] Blake, *op. cit.*, tells this whole story in a most engaging way. See Chaps. 6, 7, and 8.

solution. The Legislature in 1905 created a Board of Water Supply for the City (Chap. 724, Laws of 1905), and vested it with broad powers to find and develop new sources of water. The Board attacked its problem energetically, and shortly had concluded to go to the Catskills for the necessary new source. Suiting action to conviction, the Board in 1907 began construction of the Catskill system. The first stage, completed in 1917, consisted of the Ashokan reservoir on Esopus Creek, together with an aqueduct to carry water to the City 120 miles away. Ten years later the second stage was completed through construction of the Schoharie reservoir, whose water also came to the City through the Catskill aqueduct. The Catskill system was expensive, but it was also productive of good results. Its two reservoirs did little more than meet increasing needs, however, and it was clear well before completion of the project that further new sources would soon be required.

1949: The Bottom of the Barrel

So far the City had confined its search for new supplies to intrastate waters, hence no problem had arisen which could not be handled by state or local action. Here, however, a new phase began. Further surveys indicated the headwaters of the Delaware River as a likely new source, and the Delaware is an interstate stream, with New Jersey and Pennsylvania (and possibly Delaware) necessarily involved in any important decision concerning it. The New York State Water Supply Commission (created in 1905) immediately grasped the nature of the problem, and took cognizance of it in an early report. "It is evident," the Commission remarked, "that the most thoroughgoing development and control of this river can be secured only through some plan which shall bring the three States into effective co-operation." [3] In harmony with this conclusion, the Commission in 1908 invited its opposite number agencies from New Jersey and Pennsylvania to a conference, which discussed ". . . the possibility of co-operation between the three States in the conservation and development of the Delaware and other interstate rivers." [4] Company was good and conversation was pleasant, but no plan of action emerged from the meeting.

Fifteen years later the three states had another go at the problem. The New York City Board of Water Supply in 1921 had designated

[3] State Water Supply Commission, *Fourth Annual Report, 1909*, p. 208.
[4] *Ibid.*, pp. 208-9.

the upper Delaware and its tributaries as the next new source of water for the city. New Jersey and Pennsylvania refused to give their consent to the proposed diversion, and the three states sought agreement through an interstate compact allocating the disputed waters among them. New York was quick to approve the compact (in 1925), but the other two states refused to sanction it. A revised compact again received the approval of New York alone (1927). To complete the chapter on interstate relations, the three states (plus Delaware) in 1936 were successful in establishing an Interstate Commission on the Delaware River Basin. "Incodel," as it came quickly to be called, achieved some success in certain directions, principally in the area of pollution abatement; but its plan for allocation of the waters of the Delaware failed of adoption when, in 1953, Pennsylvania again refused to give its sanction notwithstanding earlier approval by Delaware, New Jersey, and New York.[5]

Meanwhile the Board of Water Supply, wearying of endless discussions and faced with the responsibility for meeting a need for new water which loomed ever closer, concluded to take unilateral action. It recommended, and on January 12, 1928, the Board of Estimate of New York City approved, plans for an intricate system based on a series of reservoirs on Rondout Creek (a tributary of the Hudson) and the headwater streams of the Delaware River. On May 25, 1929, the New York State Water Power and Control Commission approved plans for the Delaware project, which was to be built in three stages. Stage 1 called for reservoirs on the Rondout and the Neversink (the latter a tributary of the Delaware), Stage 2 for a reservoir on the East Branch of the Delaware, and Stage 3 for three reservoirs on upper tributaries of the Delaware. On completion of the several reservoirs, water would be conveyed by three tributary tunnels to the Delaware aqueduct, which would bring it to the City.

On May 13, 1929, New Jersey sought to enjoin the City and the State of New York from proceeding with their announced plan. The resulting case is known as the Delaware Diversion Case or the Delaware River Case. It is one of the most famous in the history of interstate stream litigation. It was here that Mr. Justice Holmes wrote the passage which begins, "A river is more than an amenity, it is treasure."[6] The City sought authority to divert 600 million gallons of water per day (mgd) from the Delaware River and its tributaries located in

---

[5] The best account of these negotiations is to be found in a mimeographed study by Maynard M. Hufschmidt, *The Interstate Commission on the Delaware Basin* (Cambridge, Massachusetts, 1956).

[6] *New Jersey* v. *New York et al.*, 283 U.S. 336 (1931).

New York State. It proposed to operate its reservoirs so as to impound water from a particular stream only when the flow exceeded the stream's ordinary flow. The Supreme Court denied the injunction sought by New Jersey, but reduced the amount sought to be diverted by New York City from 600 mgd to 440 mgd. The Court also laid down certain requirements regarding releases of water and the construction of sewage facilities. Further, in a very important proviso, it retained jurisdiction. The City had won the lawsuit, although the amount of water requested had been reduced more than 25 per cent. Even so, the amount allowed would permit construction to begin on Stages 1 and 2.

Unhappily there proved to be more to getting the Delaware project under way than obtaining legal permission to proceed. By 1931 (when permission was granted) the depression was in full swing, and several years were required to work out the problem of financing the enterprise. Shortly after work began in earnest the war came on, and activities ground to a stop. Thus it occurred that the three reservoirs of Stages 1 and 2, originally scheduled for completion in 1945, actually were not ready to receive water until 1951; and it was not until 1956 that the reservoirs were filled.

Meanwhile the City's needs crept closer and closer to the limits of its supplies, until what had appeared to be a comfortable margin of safety in the thirties looked very skimpy indeed ten years later. In 1946, consumption passed the 1,000 mgd mark, which was considered to be the safety line in terms of resources. Abnormal rainfall came to the rescue each year for three years, but in 1949 the rainfall was below normal and the reservoirs fell to critically low levels. The City took drastic water conservation measures: a rigorous inspection system was instituted against waste, public fountains were turned off, all unnecessary use of water was forbidden, and shaveless (and even bathless) days were ordained. Somehow the City lasted out the summer and fall, and good rains came to its rescue early in 1950. The call, however, had been much too close for comfort. The lesson of 1949 was well learned: it was that, even allowing for completion of Stages 1 and 2 of the Delaware project (which was still some distance away), there was need for other large new sources of water.

Alternative Additional Sources

The Board of Water Supply was, of course, keenly alert to these ominous developments. Realizing well that existing resources were far from adequate and that the new supply promised by Stages 1 and

2 of the Delaware project would not become available for some time, the Board made a recommendation to the Board of Estimate which called for vigorous action. On December 2, 1947, the President of the Board of Water Supply, in a letter to the Chairman of the Board of Estimate, suggested that the latter pass a resolution directing the former to ". . . undertake forthwith studies to ascertain what sources of supply may be quickly and economically developed to augment the present water resources of the City of New York. . . ." With the letter went a suggested resolution, which the Board of Estimate adopted on December 18, 1947. The resolution instructed the Board of Water Supply to conduct thorough studies of the future water needs of the City, and ". . . to ascertain the most available, desirable and best sources for . . . additional water supply."

In pursuance of the resolution, the water board created a Board of Consultants and directed it to survey possible alternative sources. The latter made the rounds once more, traversing what was for the most part familiar territory. Several streams east of the Hudson River were rejected as being either inadequate in flow or too remote. Certain streams tributary to the Hudson on the west were passed by on the grounds of poor quality and small yields. The Hudson itself was rejected for a number of reasons which may be better examined in another connection. The board considered tapping the Susquehanna River, but resolved against that course for reasons of both economy and water quality. A proposal to bring water from the Great Lakes was rejected because of cost and because the consent of Canada would be required. The Adirondacks, often suggested as a source in the past, were re-examined and dismissed on the grounds of cost and impracticability.[7]

Almost inevitably, then, the Board of Consultants came back to the Delaware, which had been approved for further development as Stage 3 of the plan authorized in 1929. Stage 3 had undergone some modification since its original adoption, with opinion settling over the years in favor of a single reservoir on the West Branch of the Delaware; and it was to this alternative that the Board ultimately turned. There had been some question in times past concerning the exact location of the dam which more than once had been proposed for the West Branch. The site determined upon by the Board of Consultants was located seven miles northeast of the Village of Deposit in Delaware

[7] Compilation of Studies to December 1, 1949, leading to recommendation of Third Stage of Delaware Supply Project, to Board of Consultants by Cannonsville, Division Engineer (typed report).

County. As early as 1908 it had been designated as the location of a river regulating dam by the State Water Supply Commission, which, associating the site with a hamlet in the neighborhood, labeled its proposal "the Cannonsville project." It had been found feasible for development by the Army Engineers in 1946, and had been selected by Incodel's consulting engineers as a suitable location for a storage reservoir. It was the Cannonsville site on the West Branch of the Delaware River which the Board of Consultants recommended to the Board of Water Supply in the fall of 1949.

The City's Decision

On December 28, 1949, at the very end of the dry year, the Board of Water Supply made a two-fold recommendation to the Board of Estimate:

1. that the Hudson River be tapped at a point near Chelsea for a temporary supply of approximately 100 million gallons per day to tide the City over its current emergency; and

2. that the Board of Water Supply be directed to proceed forthwith to the implementation of Stage 3 of the Delaware project, with Cannonsville designated as the site for the dam. On January 26, 1950, the Board of Estimate conducted a hearing on the Cannonsville proposal. Later the same day it passed a resolution which, referring to the plan which had been submitted to it, concluded to the effect that

> . . . the Board of Water Supply, for and on behalf of the City of New York, be and the same hereby is directed to make application to the Water Power and Control Commission of the State of New York for approval of the said map or plan of the proposed additional supply of water to meet the future needs of the City of New York . . .

and that

> . . . the Board of Water Supply . . . be and it hereby is authorized and directed to take such other and further steps as it may deem necessary in order to secure the prompt and efficient execution of the plan and project hereby approved and adopted.

On January 27, 1950, the Board of Water Supply filed its plan for the Cannonsville project with the State Water Power and Control Commission, as it had been directed to do. Its petition for approval was given the name and number of Application No. 2005, by which designation it became known among those interested in Stage 3 of the Delaware water project.

THE CANNONSVILLE HEARINGS

The Water Power and Control Commission

The Commission consists of the Conservation Commissioner, the Superintendent of Public Works, and the Attorney General, all serving *ex officio* (see above, Chapter III). Its functions are many, and it is without question one of the most important as well as one of the most active of the State's water resource agencies. It was primarily with the Water Power and Control Commission that New York City's Board of Water Supply dealt in seeking state approval of its plan for new water sources for the City. It was the Commission which conducted the hearings that are the center of interest in the present discussion.

The responsibility of the Water Power and Control Commission for water supply is indicated in the statutory provision that no municipal corporation, and no board or commission acting for such municipal corporation, shall take action for ". . . any new or additional sources of water supply . . ." until such corporation (or board or commission acting for it)

> . . . has first submitted the maps, plans, and profiles therefor to the commission, as hereinafter provided, and until the commission shall have approved the same, or approved the same with such modification as it may determine to be necessary as hereafter provided.[8]

The law provides further (in Section 523) that, on receipt of an application for approval of maps, plans, and profiles for a new or additional water supply, the Commission

> . . . shall thereupon cause public notice to be given that on a day therein named it will hold a public hearing at such place as it may specify in said notice, for the purpose of hearing all persons, municipal corporations or other civil divisions of the state that may be affected thereby.

Any person, individual or corporate, may file with the Commission written objections to the project proposed in the application, which statement must specify the ground for the objections. On the day set in the notice (and on subsequent days if more than one day prove necessary), the Commission is directed ". . . to examine the said maps and profiles and to hear the proofs and arguments submitted in support of and in opposition to the proposed project, . . ."

It is the responsibility of the Commission to determine whether:

[8] *Conservation Law*, Article XI, Section 521.

1. the project proposed is justified by public necessity;
2. the plans provide for "the proper and safe construction of all work connected with" the project;
3. they provide for the proper protection of the water supply and the watershed;
4. the proposal is just and equitable to the other public bodies and to the inhabitants affected thereby;
5. the plans make proper provision for the determination and payment of all damages which will result from execution of the project. The Commission is required to take action on the application within 90 days after the hearing. It may (1) approve the application as presented, (2) approve it with such modifications as it may deem appropriate, or (3) reject it entirely. The agency is instructed to keep a complete record of its decision and of all documents relating to the procedure. The action taken on any application is subject to review by the courts by certiorari proceedings.

In its discharge of the responsibilities above summarized, the Water Power and Control Commission conducted public hearings on March 22 and 23, April 25 and 26, June 5-7, and July 18 and 19, 1950. The full Commission (with an alternate sitting for the Attorney General) attended only the first day's hearing. All hearings were conducted at the village hall of the Village of Delhi, in Delaware County, with John C. Thompson, Secretary to the Commission, serving as chairman. The procedures were long and arduous, and the exchanges were frequently brisk, not to say sometimes heated; but the official atmosphere which prevailed was friendly, if judicious, throughout. Along in the fall, following the announcement of the Commission's decision, one of the local objectors wrote a personal letter to Mr. Thompson in which she said, in part: "My purpose in writing to you is to tell you that every hearing you conducted at Delhi last spring and summer was both educational and enjoyable to me. . . . The courtesy you extended me as well as the other objectors I shall never forget. Although I cannot approve of the decision, I feel you were fair in your treatment of everyone attending the hearings." [9]

[9] Water Power and Control Commission, The Stenographic Record in the Matter of the Application of the City of New York for Approval of Its Maps, Profile and Plan for Securing an Additional Supply of Water from the West Branch of the Delaware River, Public Hearing in the Village Hall of Delhi, March 22, 23; April 25, 26; June 5, 6, 7; July 18, 19, 1950. Without access to this official record of the hearings and the documents pertinent to it, it would have been quite impossible to arrive at an understanding of the issues involved, much less of the spirit in which the hearings were conducted.

Resistance

Anyone familiar with the beginnings of the Croton project more than a hundred years ago could have predicted that resistance would develop to the proposed Cannonsville plan. A foretaste of what was to come was provided by the appearance of William S. Fancher, attorney, at the hearing conducted by the Board of Estimate on January 26, 1950. Mr. Fancher, representing the Towns of Tompkins and Walton and the Rock Royal Cooperative Creamery, all located in Delaware County, filed a vigorous objection on behalf of his clients, emphasizing the disadvantages of the proposed reservoir to the milk producers in the area. If his protest did not gain favorable consideration, it at least served notice of a local lack of enthusiasm for the project.

Actually, the first important expression of resistance came from a non-local source. The Interstate Commission on the Delaware River Basin had interested itself in the problem of water allocation almost from its inception, and in July of 1949 had contracted with a group of consulting engineers to make a study of the problem of proper utilization of the waters of the Delaware. The engineers made a preliminary report to Incodel on January 20, 1950, and a final report on August 17 of the same year.[10] As New York City's Delaware project proposal moved into the decisive stage, there was, therefore, a great deal of discussion of the forthcoming Incodel plan. It was known that the Incodel engineers probably would propose a dam at Cannonsville as one unit in their integrated system.

Both the City and the State of New York were much interested in the prospective Incodel development. Both had been concerned for many years, indeed, in working out an interstate approach to utilization of the waters of the Delaware; it will be recalled that the State more than once had given its approval to a proposed interstate arrangement. As a matter of fact, it was a party to the agreement under which Incodel launched its engineering study of 1949. The State therefore not only was not disposed to short-circuit an interstate agreement, but on the contrary was anxious that such an agreement be reached without further delay.

The Board of Water Supply of New York City was, of course, thoroughly familiar with these developments. The President of the Board,

[10] Malcolm Pirnie Engineers, and Albright and Friel, Inc., Consulting Engineers, *Report on the Utilization of the Waters of the Delaware River Basin* (August, 1950).

anticipating some confusion from the parallel developments, wrote to the Chairman of the Board of Estimate on December 28, 1949, to this effect: "If the joint (Incodel) project becomes a reality, New York City can and should offer its Cannonsville Reservoir to the agency created by the three states to develop the Delaware River Basin for municipal water supply, subject to financial and other appropriate considerations." The temper of some of the parties to the procedure may be judged from a telephone call from the New York State Attorney General to the President of the Board of Water Supply on February 1, 1950. The Attorney General stated that some of those supporting the Incodel plan felt that New York City was "jumping the gun" with its Cannonsville proposal. Major Irving Huie (the Board of Water Supply President) wrote the Attorney General that very day setting forth the City's attitude regarding the Incodel proposal. The position he took was essentially the same as that described in his letter of the preceding December 28. Major Huie was among those who appeared as a witness at the hearings conducted by the Commission at Delhi. There he reviewed once more the policy of the Board of Water Supply, pledging his agency's complete cooperation in event the Incodel plan should become a reality. What the City sought was insurance in the form of a program of its own in the quite likely event that the Incodel proposal should fail to command the necessary approval—as, ultimately, it did fail to do. The Board by this time had had considerable experience both with abortive interstate cooperation and with droughts.

As was to be expected, there was vociferous objection locally on the part of the individuals whose lives and properties would be affected by the reservoir. A farmer and his wife joined to write a simple but very telling letter on the subject of personal dislocation. It read, in full: "Gentlemen, We hope you realize, that it takes years of life savings, to have finally a property and home and after you established same and are settled, you suddenly don't want to move. Therefore we definitely object to any water dam proposal." Another local resident wrote:

> It is kind of hard for you people from the City to appreciate what a farm means to a person who is a farmer. In the City you move from one apartment to the next and maybe they are about the same. . . . There is as much difference between one farm and the next farm as there is between black and white. . . . The general welfare of the people is one thing and it is brought up quite often. At the same time, on the other side, it says there is a right to enjoy our free life for a person and the

right to live their life. Well, then, you say "general welfare" is it any-
body's welfare to drive people out of their homes? I can't see it.[11]

Still another, a dairy farmer, testified that he had a farm of 600 acres
that would be partly flooded and wholly isolated through inundation
both of his present access road and of all practicable rights-of-way for
new roads. Another still had heard rumors of mistreatment (mainly in
the form of inequitable and delayed payments) of property owners
at the Pepacton reservoir site, and expressed doubt that he and his
neighbors around Cannonsville would fare any better.

The most original testimony was offered by Mr. Pierre DeNio, who
had no desire whatever to leave the Delaware County home in which
he had lived all his life. He described a vision he had seen. Standing
on the bank of the Delaware and looking upstream, he reported, as he
had so often done, he had observed an object in the distance moving
in the river. As the object came closer, he continued,

> . . . I saw it was a long string of birch bark canoes and in those
> canoes were a lot of ghostly figures with blankets around them, and
> feathers on their heads, and so on. They came along by where I was
> standing and looked at me and (one of them), I don't know whether it
> was Tecumseh or Powhatan . . . stood up and said, "What's the
> matter, brother?" I said, "I have got to move," and he said, "You might
> as well get your satchel and come on with us." Oh, how they did laugh.
> They laughed their heads off at me. I suppose they were thinking about
> a couple of hundred years ago.

Mr. Frank Dolan, representing the Broome County Sportsmen Association and the Broome County Federation of Sportsmen Clubs,
attacked the proposal on the ground that the concentration of so many
dams in a small area was not good public policy. He even introduced
the national defense symbol in support of his argument. In the end,
however, he came to the nub of the issue, as he saw it: his clubs pre-
ferred the streams in their natural state, and so opposed the dam.

One of the most vigorous objections was filed by a spokesman of the
Rock Royal Creamery, who maintained that, of the 153 milk producers
using the Rock Royal, 50 would be flooded out and others would have
to go to a new creamery. The district's dairy industry would be ruined,
and by a scheme that was ". . . inequitable, unconscionable, unjust,

---

[11] The law requires that a person who desires to be recognized at a public hearing must first have communicated with the Commission and presented his views in
writing. Hence these letters and other statements, which appear as a part of the
stenographic record of the hearings.

unnecessary, and presumptuous." Individual dairy farmers testified in support of this view.

A second group of local objectors included representatives of Delaware County and of the several towns and villages that would be affected by the development. Delaware County, in a resolution adopted by its Board of Supervisors on March 13, 1950, noted that there were five dams in various stages of planning or development in the country. These projects would inevitably change the country from a prosperous agricultural region to one of hills and lakes, with the best farm lands inundated in the beds of the reservoirs. The Cannonsville dam therefore would inflict grave damage on the economic and social life of the county, and further inevitably would have its effect on the tax rate. The dangers of removing the best farm lands of the county from the tax rolls were pointed out more than once. The Village of Delhi had no sewage disposal system, and at the time needed none; it wanted to make very certain that any future expense required to protect the watershed from contamination would be borne by New York City. The Town of Walton testified that it had fallen heir to a good many evacuees from the Pepacton reservoir, with consequences not entirely beneficial. It feared that the Cannonsville dam would further aggravate these unhappy consequences, which were identified as unemployment, over-crowding of schools, housing shortages, inflation of real estate values, and general dislocation.

Some of the local objectors went to considerable trouble and expense to marshal technical arguments against the proposal. For example, an attorney who represented two or three of the towns and villages produced a letter from the Commissioner of Water and Heat of Cleveland (Ohio), who testified that large amounts of water could be saved through metering. Another lawyer made the point that the Delaware development would not solve New York City's water problem permanently, and maintained that the City ought therefore to seek a longer-range solution than that promised by the present stop-gap plan. Two of the towns and a number of individual citizens went farther: they joined to invite an engineer to testify, in an effort to demonstrate that the Hudson was a better and cheaper source of water for the City than the Delaware.

Those who opposed the Cannonsville project on technical grounds had considerable ammunition at their disposal. Of the three "permanent" solutions proposed for the City's water supply problem, two (the Great Lakes and the Adirondacks) were brushed aside quickly in favor of concentration on the third. The Hudson River for many years had been regarded as an important potential source, and indeed the City even at that time had authorization to draw a modest amount of water from the river as an emergency measure. Much of the testimony taken by the Commission at its public hearings concerned the relative advantages of the Hudson and the Delaware as a source for New York City water, notwithstanding subsequent statements by representatives of the Commission that the Hudson was not officially under consideration at all.

The engineer retained by the local interests referred to above was Lawrence T. Beck of New York City, who on his own responsibility had made a study of the Hudson River. The Beck study favored the Hudson over the Delaware, and Mr. Beck himself testified vigorously in support of the Hudson. Further, his plan commended itself to the Citizens' Budget Commission of New York, which joined in its support. During the course of his presentation, Beck testified that he had discussed his plans with two members of the staff of the Corps of Engineers of the New York District office. Lest there be some suggestion of official Corps approval in this testimony, Colonel Edwin P. Ketchum, District Engineer, wrote a letter to the Board of Water Supply on May 16, 1950, disavowing any intent to voice formal endorsement of the Beck plan. Harold Riegelman, Counsel to the Budget Commission, testified further in behalf of the Hudson River plan in a letter to the Water Power and Control Commission dated October 4, 1950. His letter emphasized the favorable cost-per-million-gallons position of the Hudson River over the Delaware proposal, particularly in the event that the latter should become part of the larger Incodel scheme.

Whether or not the Hudson River plan was officially before the Commission, the Board of Water Supply felt obliged to rebut the testimony which had been introduced. This it did first by disputing the comparative cost figures introduced by those who favored the Hudson. The Board found the anticipated capital cost of the Cannonsville project to be about $140,000,000, that of a Hudson project to be built in stages to yield the same amount of water close to $85,000,000. To the latter figure must be added an amount, not calculable, to reflect the higher construction costs likely to prevail at the time of future additions to the project. Cannonsville's annual cost, the Board continued, would decrease from about $7,500,000 to about $4,500,000 over a forty-year amortization period, and thereafter would be around $500,000. The Hudson project, by contrast, would increase from $4,300,000 to as much as $9,600,000 a year, and thereafter would continue indefinitely at about $7,100,000 a year. From its computations the Board concluded that Cannonsville enjoyed a substantial cost advantage over the Hudson. Further, the Board argued, there would be no comparison between the two in the matter of cost and inconvenience to the respective property owners. On June 1, 1950, the Chief of the Bureau of Claims of the Board of Water Supply submitted a report to the President of the Board on the effects of a barrage dam in the Hudson (the central feature of the Beck plan) on the properties involved. The report included letters from a hundred or so companies and municipalities that would be affected by the resulting rise in the water level of the river. It purported to show that industrial property, shippers, towing companies, operators of recreational facilities, owners of private dock and boating facilities, municipalities along the river, the State of New York, and New York City all would suffer damage from such a dam. The chief engineer of the New York Central Railroad confirmed this testimony in part, estimating that it would cost his company $200,000,000 to raise its main line three feet from Chelsea to Albany, an operation which would be required if the Beck plan were adopted.

In the end, the Board of Water Supply rested its objection to the Hudson River plan on considerations of water quality. "Regardless of cost," said the Board in a report to the Mayor, "the paramount factor influencing our conclusion is that the quality of the water which Mr. Beck would supply to the people of New York City from the Hudson River is definitely inferior and its use is . . . hazardous." [12]

The Case for Cannonsville

Ultimately, the Commission's decision would of course rest on the strength of the Cannonsville plan. The Board realized this well, and while it reacted to all criticisms and adverse claims by counter-argument, it rested its case primarily on an affirmative presentation. The main considerations in support of the Cannonsville plan were these:

1. The water available from the Delaware ("upland water") was held to be of higher quality than any other available in like quantity and at comparable cost.

2. Water from the Delaware could be had for less money, in terms both of original outlay and operating cost, than could water from any other available source.

[12] *New York Times,* July 28, 1950, p. 1, col. 2.

3. The Cannonsville reservoir had been planned as Stage 3 in the unified development of the upper Delaware. As part of that plan, the Delaware aqueduct, which would carry water from all the reservoirs to the city, had already been completed.

4. Water would flow from the Cannonsville reservoir to the city by gravity, thus eliminating the need for pumping.

5. Cannonsville could be financed by borrowing against the improvement outside the City's debt limit.

6. Notwithstanding individual cases of loss and inconvenience, construction at Cannonsville would involve less economic and social loss than would development at any other site available.

In a memorandum filed with the Commission on August 15, 1950, the City summarized its case. Addressing the chief criteria for judgment named by law, the City undertook a point-by-point examination of its case. Would construction on the project be safe and proper? The City pledged that all work would be undertaken in accordance with proven methods and designs, plans for which would, of course, be submitted to the Commission for prior approval. Was the proposed plan just and equitable to other municipal corporations and civil divisions? There was no other municipality which was so located as to require any considerable amount of water from the West Branch, yet New York would be obliged to accommodate any such need should one arise. Was the City able and willing to settle all claims equitably and promptly? It affirmed that it was and undertook that it would do so, citing its past record to prove both its capacity and its good intention. The memorandum revealed an undertone of satisfaction; it was clear the City felt it had made its case.

Administrative Considerations

With the testimony of the partisans in, there yet remained some loose ends for the Commission to tie up. In particular, there were certain state departments which had collateral interests in a water supply application. Among these was the Conservation Department, for while the Water Power and Control Commission itself headed a division within that department, there were other divisions which might be expected to have an opinion to express. Forehanded, the Board of Water Supply had already made an agreement with the Department which took account of the fact that ". . . the areas affected include some of the State's most important fishing waters. [This is] most important, because they provide fishing opportunity within easy reach of the metropolitan area, where the demand is enormous." [13] The agreement pledged the city, ". . . in the interest of the propagation, conservation and preservation of fish life . . . ," to release enough water to maintain a brisk minimum flow in the stream below the dam. Thus was one official hurdle cleared, on the initiative of the City. In addition, the City agreed at its own expense to construct barrier dams on tributary streams at sites to be agreed upon with representatives of the Conservation Department and on property owned or to be acquired by the City.

The Department of Health also had an important interest in the reservoir. In a letter (undated) to the Water Power and Control Commission the Commissioner of Health gave voice to that interest and made a recommendation regarding it:

> (a) That this application (Application No. 2005) be approved insofar as sanitary features may be concerned with the requirements that rules and regulations be enacted and enforced by New York City under the provisions of Article V of the Public Health Law for the protection of the proposed supply from the West Branch of the Delaware River.
>
> (b) That any such approval be contingent upon the effective chlorination of water to be secured from the proposed reservoir.

Major Huie vouchsafed the desire of the Board of Water Supply to cooperate fully with the Health Department. If the Department (through the Water Pollution Control Board) should conclude that additional purification was necessary, Mr. Huie said, the City would expect to bear the cost. In response to a question by a member of the Commission, Charles R. Cox, representing the Department of Health, said that later on representatives of his department and the City would get together ". . . to develop consistent rules which would be workable without being too rigid or too lenient. . . ."

While the record does not reveal it, there likely were negotiations between the Board of Water Supply and the State Department of Public Works regarding the re-location of state highways in the reservoir area. In any event, a letter dated June 7, 1957, from the Deputy Chief Engineer of the Division of Construction of that Department related that plans for highway re-location had been developed by Board engineers and approved by the Department. The Department will authorize construction of the new highways, the letter concluded,

[13] Cecil E. Heacox, "Liquid Assets," reprint No. 113 from the *New York State Conservationist* (1950).

with funds deposited with the State Comptroller by the City of New York.

The public hearings concluded, the Water Power and Control Commission retired to study the record. It continued to receive communications during the succeeding weeks, but its file on the application was essentially complete by mid-summer. The Commission's decision was made public November 14, 1950.

The document announcing the decision contained three basic parts.[14] First, there were 54 "Findings of Fact," which may be briefly summarized thus:

1. Physical features of the project: A number of findings described the project in some detail.

2. Objections: The objections were analyzed with care, and were rejected one by one. The Commission expressed its sympathy for the objectors, particularly for the local residents who would be dispossessed, but found insufficient reason in the opposition arguments for denying the application.

3. Alternative sources: The Commission examined the proposed alternative sources, and dismissed each in turn. It was pointed in its rejection of the Hudson River proposal (the Beck plan), on which it looked with strong disfavor.

4. The Cannonsville proposal: The Commission found that the City had conducted a thorough search for new sources of water, that it had settled upon the Delaware River with good cause, that it had presented a workmanlike plan, and that the experience of the City had been such as to warrant the belief that it could and would carry the project through to conclusion in compliance with statutory and administrative requirements.

The second major section of the Decision document contained a series of eleven conditions which the Commission found it necessary to lay, notwithstanding its generally favorable attitude toward the application. Most of the conditions were calculated to ensure compliance either with statutory requirements or with the regulations of other state administrative agencies. Two or three are worthy of special mention. One provided that the Commission's decision should have no effect unless and until the Supreme Court of the United States had

[14] Water Power and Control Commission, *Water Supply Application No. 2005, Decision* (November 14, 1950), mimeographed.

modified its earlier decree to give the City authority to take additional water from the Delaware. Another specifically reserved the right of the Commission to pass on detailed plans and specifications prior to the beginning of actual construction. Still another directed attention to the fact that no part of the works could be operated until they had been approved by the Commission as completed. "In general," the reservation concluded, "such approval will not be given except for a fully completed system, and it will never be given until all provisions affecting quality of the water and safety of the works fully have been complied with."

Finally came the "Statutory Determinations" and the decision itself. These follow:

> The Water Power and Control Commission, having given due consideration to the said petition and its exhibits, and the proofs and arguments submitted at the hearing determines and decides as follows:
>
> First. That the application, maps and plans submitted are modified as set forth above and, as so modified, are the plans hereinafter mentioned.
>
> Second. That the plans proposed are justified by public necessity.
>
> Third. That said plans provide for the proper and safe construction of all work connected therewith.
>
> Fourth. That said plans provide for the proper protection of the supply and the watershed from contamination and that filtration is at the present time unnecessary.
>
> Fifth. That said plans are just and equitable to the other municipalities and civil divisions of the State affected thereby and to the inhabitants thereof, particular consideration being given to their present and future necessities for sources of water supply.
>
> Sixth. That said plans make fair and equitable provisions for the determination and payment of any and all legal damages to persons and property, both direct and indirect, which will result from the execution of said plans or the acquiring of said lands.
>
> WHEREFORE, the Water Power and Control Commission does hereby approve the said application, maps and plans of the City of New York as thus modified.

The decision was signed and sealed on November 14, 1950.

The project which the Commission had approved consisted of two major features. First, the reservoir would form behind an earth-fill dam approximately 170 feet in height above the stream bed. The dam would impound about 97 billion gallons (298,000 acre feet, 13 billion cubic feet) of water at a flow line elevation of 1150 feet above mean sea level. The reservoir would have a watershed drainage area of about 450 square miles, and would yield an estimated 388 million gallons of water per day. Total costs appurtenant to the dam were estimated to run to $59,297,000. Second, the project called for a tunnel 10 feet in diameter to link Cannonsville with the Rondout reservoir, through which the water would have access to the Delaware aqueduct. The tunnel would be 44 miles in length, and would cost an estimated $80,714,000. The total cost of the reservoir and tunnel therefore would come to somewhat more than $140,000,000. The project was estimated to require ten years for completion.

On April 1, 1952, the City petitioned the Supreme Court to reopen the 1931 case and increase the allowable diversion from the Delaware basin from 440 to 800 million gallons per day. On June 7, 1954, the Supreme Court took favorable action on the petition, entering a decree permitting diversion by the City of 800 million gallons a day total.[15] Shortly thereafter the City submitted detailed plans to the Water Power and Control Commission, as it had been directed to do. The project actually got under way with ground-breaking ceremonies on December 19, 1955. The Board of Water Supply reports, with understandable satisfaction, that the local residents have cooperated well in the business of acquiring title to property. No appreciable difficulty has been encountered on this potentially troublesome front.[16]

Fact No. 29 developed by the Water Power and Control Commission reads thus:

> It is estimated that the amount of water asked for by the present project will, together with the amount secured from existing sources and from those being presently developed, be in the neighborhood of 1,800 million gallons daily, an amount which is expected to be sufficient to meet present and future needs of the city to at least the year 1980.

Those familiar with the history of the City will know that its officials for more than a hundred years have spent a good share of their time attending ground-breaking ceremonies heralding the beginning of new water projects and fireworks displays celebrating their completion, and will wonder how sound this prediction will prove to be. Others, not necessarily readers of history, will wonder, too. Indeed the whole Delaware project is regarded by some as nothing more than a chapter in the City's unceasing quest for water. This applies particularly to Cannonsville. This trend of thought is illustrated by an exchange which occurred during the public hearing at Delhi. The parties to the col-

---

[15] U.S. Supreme Court Decree, 347 U.S. 995.

[16] Board of Water Supply of the City of New York, *Fiftieth Annual Report,* 1956.

loquy were Mr. Fancher, attorney for the Towns of Walton and Tompkins; Mr. Thornton, Assistant Counsel for the Board of Water Supply; Mr. Thompson, presiding over the hearings for the Commission; and Mr. Fitzgerald, Chief Engineer for the Board. The record on the point reads:

Mr. Fancher: "Where are you going after you get through with Cannonsville, Mr. Fitzgerald?"

Mr. Thornton: "I object to that."

Mr. Thompson: "Do you know where you are going after you get through with Cannonsville?"

Mr. Fitzgerald: "No, sir." (Laughter.)

If Mr. Fitzgerald does not know where the City will go next for its water, then probably no one knows. But if the direction cannot be foretold, the fact of movement can; for notwithstanding the optimistic forecast for the next quarter-century, past events warrant the expectation that New York City will be on the search for new water sources at a not-so-distant date. There will always be an asphalt desert.

CHAPTER VI

River Control: The Black River War

By all odds the most exciting chapter in New York's none-too-proud history of river control is that written around the efforts of the Black River Regulating District to construct a storage reservoir in the Adirondacks. It is the story of a semi-autonomous local unit, set up as a receptacle of state authority and responsibility, which suffered frustration after frustration along its road to eventual defeat. In particular, it is the story of the metamorphosis of a local problem, so defined by law, into a statewide political issue, with a consequent decision based on considerations only indirectly related to the original question. The events of the Black River War shed light on a number of points important to the field of water management.

THE RIVER REGULATING DISTRICT

The background of the State's experience in river control has already been examined at length.[1] It is necessary now to sketch in with somewhat more care the legal framework for river regulation. The development of river regulation as a concept and that of the Forest Preserve were intimately related.

The "forever wild" principle, which had been written into the statute creating the Forest Preserve in 1885 (Chap. 283, Laws of 1885), gained added status with its inclusion as Section 7 of Article VII of the Constitution of 1894. The policy was complicated neither in concept nor in statement: "The lands of the state . . . constituting the forest preserve . . . shall be forever kept wild as forest lands." The intent of the Constitution was clear enough; it was to safeguard the grandeur of the forest, to protect the newly established reserve from invasion by "de-

[1] In Chapter II, especially in the section on Water Control.

velopers," whether public or private. Put forward tentatively and without full realization of its implications in the beginning, the principle has gained acceptance beyond any reasonable expectation. "Forever wild" has become the battle cry of conservationists of every hue and description. Around this talisman have rallied forces highly influential in the gradual accumulation of New York's great Forest Preserve, which now encompasses more than 2,400,000 acres in the Adirondacks and the Catskills.

Shortly after the forever wild concept achieved constitutional status, an unanticipated dilemma regarding water control began to take shape. First, it came to be widely accepted that there could be no real control of stream flow without water storage reservoirs; but second, many of New York's most important rivers originated in the Forest Preserve in the Adirondacks, which was barred to storage reservoirs by constitutional provision. The conflict between river control and the constitutional prohibition on reservoir construction claimed increasing attention, with a number of early state water agencies calling attention to the problem and urging that it be resolved. The resolution usually suggested involved an amendment to the constitution—the Water Supply Commission, for example, urged this step in its report of 1910.[2] The Commission had previously conducted a survey of potential reservoir sites, and had concluded that a general plan of river control by storage reservoirs would require utilization of less than 4 per cent of the Forest Preserve lands. Its proposal for constitutional amendment found favor, and in 1913 an amendment was approved which permitted

> . . . the use of not exceeding three per centum of such (Forest Preserve) lands for the construction and maintenance of reservoirs for municipal water supply, for the canals of the state and to regulate the flow of streams. . . .

This amendment gave constitutional sanction to limited encroachment on the Forest Preserve. While its practical effect was minimal in the end, it did clear the way for a restricted program of reservoir construction where the reservoirs would do the most good.

Two years later the Legislature passed a law authorizing the creation of river regulating districts.[3] The act provides that any property owner, private or municipal, may petition for the formation of a regulating

---

[2] Water Supply Commission, *Fifth Annual Report, 1910* (Albany, 1910), p. 112.

[3] Chap. 662, Laws of 1915. This law, widely known as the Machold Storage Law, is incorporated in the current Conservation Law as Article VII. The summary which follows rests on the law as it appears in Article VII.

district. The petition is reviewed by the Water Power and Control Commission, which is required to hold hearings before making a final determination. A regulating district has the status of a public corporation, and as such enjoys perpetual existence. It is given the power (Sec. 431)

> . . . to acquire and hold such real estate and other property as may be necessary, to sue and be sued, to incur contract liabilities, to exercise the right of eminent domain and of assessment and taxation and to do all acts and exercise all powers authorized by and subject to the provisions of this article.

The powers of the districts are vested in a board of three persons appointed by the Governor, two of whom must be freeholders resident in the district, for five-year terms.

It is worthy of note that the river regulating district, though instigated locally for local purposes and managed by a board resident in the district, is nevertheless a state agency. This was made clear by the Appellate Division of the New York Supreme Court, which in a well-known opinion confirmed the constitutionality of the statute creating the regulating district and held further

> . . . that the respondent district is a state agency; that the members of the board of the district are state officers and that their acts are acts of 'the state,' and comply with the statute.[4]

The board of the regulating district is directed to prepare a general plan for the regulation of the flow of the river or rivers in its district. The plan must show full detail on existing structures and on proposed new structures—location of each reservoir, number of acres to be flooded by each, status of ownership of property affected, estimated value of the land to be taken, and so on. On approval of the general plan by the board, it is then submitted to the Water Power and Control Commission, which ". . . may approve the same, or modify and approve it as so modified." The plan as approved becomes the "official plan" for the district, and may be modified thereafter only by the same procedure as that by which it was originally adopted.

For any particular reservoir, the board must cause to be prepared detailed preliminary plans, including maps, specifications, and cost estimates; and the preliminary plans, on approval by the board, must be forwarded to the Commission for approval (as submitted or modified)

---

[4] *Adirondack League Club* v. *Board of the Black River Regulating District,* 275 A.D. 618 (1949).

and certification. On approval by the Commission, the board must conduct a hearing in the district. After the hearing the board determines whether or not to proceed with its preliminary plans, either as originally adopted or as modified, and certifies its determination to the Commission, which must approve the final plans as before. The law makes provision for the customary court review of the board's final action by certiorari. Unless there is application for certiorari review within thirty days, or on disposition of such action in the event application is made, the final plans are regarded as officially adopted and the board may proceed with construction.

Construction is financed by sale of bonds, which are secured by assessments levied against the public corporations and properties benefited. Such bonds are not regarded as state obligations, but as obligations of the district alone. Once construction is completed, the board has charge of the maintenance and operation of the reservoir. Maintenance and operation costs, like the original construction costs, are levied against the public and private beneficiaries of the flow regulation.

Since the act became effective, petitions have been received by the Water Power and Control Commission requesting the creation of four regulating districts. Petitions for districts in the Raquette River and the Salmon River watersheds were withdrawn after a variety of local organizations registered vigorous protests. Petitions for two regulating districts received favorable consideration. The Black River Regulating District received approval first, and was organized in 1919. The Hudson River Regulating District was set up in 1922. It proceeded to construct the Sacandaga reservoir, which is regarded as highly successful. The Hudson River District has experienced a reasonably quiet existence, though beset by tribulations in its early days. The Black River District has not had an easy moment since the day of its creation, for the measures proposed by its board have been challenged at almost every step.

PRELIMINARY ENGAGEMENTS

The Black River drains some 1,918 square miles on the southwesterly slopes of the Adirondack plateau.[5] The River rises in the wooded highlands of Herkimer County, flows southeasterly for a short distance and

[5] The general description of the area which follows is derived largely from a pamphlet published by the Black River Regulating District titled *River Regulation in the Black River Regulating District* (Watertown, 1945).

then makes its way in a generally northwesterly direction to Lake Ontario at Black River Bay (Map 4). Its eastern tributaries, Moose River, Beaver River, Independence River, Otter Creek, and Woodhull Creek, drain from the Adirondacks. On the west, the Deer and Sugar Rivers flow from Tug Hill, which consists of a high plateau separating the Black River valley from Lake Ontario. Most of the flow above the village of Forestport is diverted southward by a storage dam through a feeder canal to the Barge Canal.

The slopes from which the Black River and its tributaries flow range in elevation up to 3,000 feet. The River descends about 320 feet in the 20 miles from Forestport to Lyons Falls, where the Black and the Moose meet. There is an abrupt 60-foot fall near the confluence, and then the River descends less than 10 feet for the next 41 miles to Carthage. This is a broad, open valley from .5 to 2 miles in width with an elevation of about 725 feet. The gently sloping flood plain, referred to as the Black River flats, is but a little higher than the river at low stage and is subject to annual flooding. From Carthage to Lake Ontario, a distance of 31 miles, the River falls 480 feet.

The River's power potentialities were recognized early, and a paper mill was established at Watertown in 1808. At the present there are 18 hydroelectric plants, 19 paper and pulp mills, and several other factories utilizing the waters in the watershed. The paper industry and farming, principally dairying, have long been and are now the economic mainstays of the 80,000 inhabitants of the region. The basin's economy, which is geared closely to the river, requires a uniform, year-around flow for maximum productivity. This boon it does not enjoy, however, for the Black River, like other Adirondack streams, is plagued by periodic floods and droughts. The cause of floods usually is the spring snow melt, though heavy rains in the summer and fall (mean annual rainfall is 46 inches) often wreak their havoc. The few reservoirs available to contain the flood waters are not adequate. As a result, when the floods are followed by summer droughts, there is little water in the stream to turn the wheels of industry. The mills must restrict their operations, and occasionally they are forced to shut down. As for the fertile acres in the flats, they become momentarily unproductive as farm land and are useful only for grazing.

The diversion of water from the Black River basin to feed the Erie Canal in 1849 aggravated these conditions, particularly with respect to summer flow at Watertown; and the State made compensation in kind by constructing three reservoirs to ameliorate low-flow conditions. Two of the reservoirs, Old Forge and Sixth Lake, were created in 1880 by raising the levels of natural lakes on the middle branch of the Moose River. The other, Stillwater reservoir, was built in 1882 on the Beaver River. The effectiveness of the reservoirs was limited, due mainly to their small capacities. It was this combination of long-felt need and ineffectual remedial action that prompted a number of municipalities and industries in the area to join in a petition to the Conservation Commission for the establishment of a river regulating district under the law of 1915.

Following the creation of the Black River Regulating District on May 7, 1919, the Board drew up a preliminary plan which envisioned the comprehensive control of the flow of the streams within its jurisdiction through a series of storage reservoirs. According to the official plan adopted March 22, 1920, the system would include twelve reservoirs plus the three already in existence. One of the latter, Stillwater reservoir, was slated to be enlarged. Even then, the contemplated system would control only 40 to 50 per cent of the River's total flow. As the Board stated, however,

> This appears to be about the practical limit of control by regulating reservoirs, because of the lack of feasible basins on most of the smaller tributary streams.[6]

Actually this degree of control would mitigate the harmful effects of floods and ensure sufficient flow to meet demands throughout the year. As it turned out, the majority of the proposed reservoirs subsequently were proved to be unjustified for one reason or another, so that in the end only the Stillwater enlargement, the Panther Mountain and Higley Mountain proposals on the south branch of the Moose, and the Hawkinsville project on the Black below Forestport remained for consideration.

The Black River War began with preliminary skirmishing in 1922, when construction plans were announced for an enlarged Stillwater reservoir. The validity of the Stillwater action was attacked in court. The New York State Supreme Court upheld the District, and the decision was affirmed by the Appellate Division.[7] The Board proceeded with the new construction, which raised the existing dam 19 feet and increased the capacity of the reservoir to 4.7 billion cubic feet (104,000

---

[6] Black River Regulating District, *Conservation of Water Resources* (Watertown, 1949), p. 33.

[7] *Board of the Black River Regulating District* v. *Wilson D. Ogsbury*, 203 A.D. 43 (1922). Affirmed by 235 N.Y. 600 (1923). This is widely regarded as a test case.

acre feet, somewhat more than 35 billion gallons). The area inundated
was increased from 2,800 acres to 6,700 acres, including 3,092 acres of
Forest Preserve land. Apart from the court action noted, no substantial
resistance appeared. As one former official of the District explained,
"There were no people around here with wealth, as there were later at
Panther Mountain and Higley Mountain. We were lucky to get down
before the furor started over state lands."

The enlarged Stillwater reservoir was a success from the moment of
its completion in 1925. Destructive floods on the Beaver River have
been completely eliminated: the ratio of high flow to low flow is now
8 to 1, which reflects an extremely equable flow for an Adirondack
stream—the still unregulated Moose (the subject of the Panther Mountain dam controversy) has experienced a high-low ratio of 50 to 1
within the last five years. The regulated flow has also helped with flood
control on the Black River, to which the Beaver contributes 9 per cent
of the total flow at Carthage. The benefits associated with recreation
and sanitation are apparent, but not measurable. Most important of all,
perhaps, the Beaver River valley has prospered. Let James Lewis,
President of a Beaver Falls paper company, state the case in his own
words:

> In the days before the new dam, they could grind pulp in only two
> seasons, spring and fall. They would keep men working all hours of the
> day to get the pulp ground while there was water. When there wasn't
> water, the men didn't work . . . Now with a steady flow, we have a
> steady operation. The people around here work all year. There are
> even new industries coming here. We generate our own power, and
> even have some to sell.[8]

The operation of the Old Forge and Sixth Lake reservoirs has not been
as successful as a regulating device. The area around both reservoirs has
been built up as a resort and recreational area. Any draw-down in the
summer months would be most unwelcome; so the District maintains
the reservoirs at a high level until after Labor Day. This severely
limits their utility for low flow supplementation.

The first move of any consequence after completion of the Stillwater
project was made by the Board in submitting an application to the
Water Power and Control Commission requesting a change in the District's official plan, ". . . eliminating therefrom the proposed Higley
Mountain reservoir and (increasing) the storage capacity of the pro-

---

[8] Interview with James Lewis, president of the J. P. Lewis Paper Company,
June 12, 1957.

posed Panther Mountain reservoir from 4.4 to 14 billion cubic feet." [9] The Commission demurred, and suggested instead that the Higley Mountain project be retained as part of the over-all plan and that the capacity of the Panther Mountain reservoir be increased, but not as much as the Board had requested. The Board agreed and the modification was approved September 30, 1932. The purpose underlying the move was to increase the capacity of the more favorable reservoir and thus arrive at a better cost-benefit ratio.

Preliminary plans and specifications in line with the revised program were compiled and submitted to the Water Power and Control Commission, which approved them on May 7, 1935.[10] The Board made its final order and determination June 24, 1935. Having complied with the procedural requirements, the Board applied in turn to several Federal relief agencies for aid in financing the project, estimated to cost about $4,000,000. Nothing came of the requests. A statement of the problem by a former district official in a 1941 memorandum to the Water Power and Control Commission is revealing. The memorandum read in part as follows:

> During the past four years the Board has made every effort to finance the larger reservoir, but up to date, without success. The owners of the properties subject to assessment have not felt able to assume so large a tax burden without some governmental assistance. An application for a grant-in-aid was made to the Public Works Administration, but, while the project was approved as to engineering and economic soundness, the amount of money available to the Administration was insufficient to permit the inclusion of the Panther Mountain project in the P.W.A. program for this area.[11]

Toward the beginning of this period the same official had been told by a P.W.A. officer, "I'll do what I can for you, but I don't think you'll get a nickel. You folks up there don't vote right."

There was, however, a Federal agency interested in the Black River area. Under provision of the Flood Control Act of 1936, the Army

---

[9] Conservation Department, Division of Water Power and Control, *Fifth Annual Report,* 1931 (Albany, 1932), p. 9.

[10] Water Power and Control Commission, *Ninth Annual Report,* 1935 (Albany, 1936), p. 10. If the procedure appears unduly dilatory, it must be remembered that these were depression years, and that the Black River Regulating District, in company with most enterprises, public and private, was watching trends and playing by ear.

[11] E. S. Cullings, "Memorandum on Higley Mountain Reservoir with Respect to Amendment No. 3, Official Plan, Black River Regulating District, June 16, 1941, to Water Power and Control Commission," p. 3.

Corps of Engineers began a survey of the Moose and Black Rivers. On October 10, 1941, a recommendation was forwarded to the Speaker of the House of Representatives favoring the construction of a multiple purpose reservoir of 14 billion cubic feet capacity at the Panther Mountain site. The estimated costs were $3,470,000, of which the regulating district had indicated a willingness to contribute $2,920,000.

In the meantime the Board, growing restless at the delay, in May 1941 had passed a resolution to amend the official plan to include a specific proposal for a Higley Mountain reservoir at an estimated cost of $1,738,000.[12] The decision was motivated by several factors. No Federal aid had been received, nor was prospect of such aid in sight. The beneficiaries were not anxious to pay the additional taxes which the larger project would have required. The lower costs of the Higley project together with an increase in the proposed reservoir's capacity to 5 billion cubic feet seemed a suitable compromise in the circumstances. On November 25, the Water Power and Control Commission approved Amendment No. 3 to the District's official plan incorporating the Higley Mountain project and providing for a Panther Mountain reservoir of 7 billion cubic feet.

Preliminary plans for the Higley Mountain reservoir, having been adopted by the Board, were approved by the Commission in March 1942. This was immediately followed by a public hearing, and this in turn by a final authorization to commence construction. All action ground to a halt at that point because of court proceedings against the final order. The plaintiff was Associated Properties, a holding company for power property of the International Paper Company. The plaintiff, a prospective beneficiary of the project, sought a delay in the assessments. Ultimately the company withdrew and the case was dismissed, but by this time it was midyear of 1943.

One feature that marked the early Higley Mountain negotiations was the support given it by those who later blocked its construction. Take, for example, the stand of the Adirondack League Club, whose lands would be flowed by either Panther Mountain or Higley Mountain reservoir. In a letter to the Secretary of the District, the President of the Club wrote:

> The Club's main interest is in the preservation of the wilderness and the protection of fish and game in it; and the Club would therefore

[12] Minutes of the Board of the Black River Regulating District, Meeting of May 6, 1941, Watertown, New York (in the files of the District). Hereafter cited as *Minutes.*

prefer not to see the wilderness encroached upon. However, we have no desire to stand in the way of necessary improvements. That part of the Moose River which flows through the Club Preserve has no water power site capable of commercial development, and the Club would therefore not expect to pay any part of the cost of the project because it would not receive any benefit therefrom.

The Club therefore would make no objections to the building of the Higley Mountain Dam on the following terms . . .[13]

By now the construction of Higley Mountain reservoir was indefinitely postponed due to the war. Nevertheless, expressions of support for it continued to come in. For instance, Lithgow Osborne, then Conservation Commissioner and later a leader of the anti-dam forces, reportedly told the District's Secretary, "Well, if you're going to build a reservoir somewhere, [Higley Mountain] seems to me an awfully good place to build it." Likewise, the Association for the Protection of the Adirondacks, a conservation group later headed by Osborne, somewhat later took a stand not opposing the Higley Mountain Reservoir.

Into the midst of these developments dropped Public Law 534 (78th Congress, 1944), approving the Panther Mountain reservoir as a Federal project. Activity relating to the Higley Mountain project was suspended pending investigation into the Panther Mountain prospects. Subsequently, on October 31, 1945, the Board resolved to provide without cost to the United States the necessary lands and easements, to maintain and operate the reservoir after its completion, and to hold the United States free from damage claims arising from construction of the reservoir. Local interests were to contribute the bulk of the cost of the project. In taking these steps, the District acted as the state agency responsible for Federal-State negotiations.

In the meantime the conservationists, galvanized into action by the threatened invasion of the wilderness, moved to consolidate their position. Prior to this time, as one of their leaders explained, conservation organizations were for the most part local in nature, with little thought of the need or the efficacy of concerted action. The reservoir controversy jarred them out of their complacency and energized them into vigorous motion. One mail, this spokesman continued, brought letters from thirty-five organizations he had never heard of enlisting in the fight for the Forest Preserve. The conservationists insist that their major concern was to maintain the forever-wild principle; the reservoirs

[13] Letter from Oscar Houston, President of the Adirondack League Club, to E. S. Cullings, April 25, 1942.

were not regarded as bad in themselves, but only as invaders of the forest.

On October 21, 1945, spokesmen of thirty clubs and leagues conducted a conservation forum. A resolution offered by a representative of the Wilderness Society of Washington, D. C., declared that ". . . construction of the reservoirs would destroy the largest deer winter-yardage in the Adirondacks, exterminate deer that seek winter food and shelter in the valley, and destroy some of the best natural brook trout waters." The meeting went on record as unanimously opposed to the construction of the reservoirs. A second resolution urged Governor Dewey and the Conservation Department to aid in preventing their construction. The Conservation Commissioner, Perry B. Duryea, who was present, promised that his department would oppose the reservoirs. The Adirondack Mountain Club of Albany was authorized to assume the leadership in forming a state-wide opposition group.[14]

The outcome of the October meeting was the formation of the Adirondack Moose River Committee, Inc., the following February. The letterhead of the organization carried this statement:

> The objective of the Adirondack Moose River Committee is to preserve the wild forest character of the valley of the South Branch of the Moose River and other similar areas in the Adirondacks now threatened with destruction by the proposed creation of unnecessary reservoirs.

The letterhead also contained a formidable list of names and groups. The organizations represented by the officers and executive board were:

Montgomery County Conservation League
Saratoga County Fish and Game Council
Schenectady County Conservation Council
Oneida County Conservation Council
Izaak Walton League of America
Adirondack Wilderness Committee
New York State Conservation Council
Forest Preserve Association of N. Y. State
Buffalo Conservation Forum
The Wilderness Society of Washington, D. C.
The Adirondack Mountain Club
The Schoharie County Conservation Alliance
The National Parks Association
Warren County Federation of Sportsmens Clubs

[14] *New York Times,* October 22, 1945, p. 17.

Wildlife League of Albany
Otsego County Sportsmens Federation
Tompkins County Fish and Game Club
Rensselaer County Conservation Alliance
Clifton Park Fish and Game Club

A move made immediately by the new group emphasized its single-mindedness. It expressed the hope that control over the use of the Forest Preserve for such things as regulating reservoirs could be vested exclusively in the Conservation Department. Such a move would end the arrangement by which the Conservation Commissioner shared authority with the Attorney General and the Superintendent of Public Works as members of the Water Power and Control Commission.[15]

On the very day the Adirondack Moose River Committee was organized, the chairman of the Legislative Assembly's Conservation Committee, Leo Lawrence, introduced a bill to block further construction of reservoirs for power generation in the Forest Preserve by river regulating districts. The bill reportedly carried the endorsement of the Conservation Department. The conservation groups fought for passage of the measure, while the Black River Regulating District recorded its vigorous opposition. The District retained a former state assemblyman, Russell Wright, to represent it in opposing the proposed legislation. The Assembly passed the bill, 81 to 44, but it was defeated in the Senate by a vote of 21 to 26.

On July 1, 1946, the Board of the Black River Regulating District resolved to undertake construction of the Higley Mountain reservoir. Although Congress had authorized the Panther Mountain project, no appropriation had been made for the Federal government's share nor was there any assurance that any would be forthcoming. The District therefore initiated contract and land condemnation proceedings for the Higley Mountain project.

In the following January, Assemblyman Lawrence drew a bead on the Black River Regulating District. His weapon was a legislative bill which was similar to that of the preceding year except that it did not include the Hudson River Regulating District. In explaining this change Lawrence said,

> The record shows that the Hudson River Regulating District has religiously conformed to the mandates of the Constitution. It has done a masterful job along the lines of flood control. . . .
> The Black River Regulating District on the other hand has failed

---

[15] *New York Times*, February 22, 1946, p. 23.

utterly and entirely to carry out the mandates prescribed in the Constitution . . . The reason being, instead of working toward ends of flood control, it has utilized its every energy to develop hydro-electric power.[16]

No "flood control mandate" has been found in the Constitution, and certainly the history of river regulation in the State affords little basis for this view. Nevertheless the Lawrence statement foreshadowed the confusion, accidental or intentional, that arose over the purpose of river regulation districts.

The sportsmen's groups converged on Albany to fight for the bill. The pro-reservoir forces waged an equally energetic campaign against it. In the end the bill was defeated in the Assembly by a vote of 71 to 63, five votes less than the absolute majority of 76 required.

In September of 1947 Governor Dewey requested that the District hold up work on the project until an investigation could be made. Two months later the Higley Mountain reservoir was scrapped by executive action. By law, only the Board of the Black River Regulating District could rescind the plan. The Governor accomplished his purpose by filling two existing vacancies on the Board. The newspaper account of the change read thus:

> The Governor said today that William R. Adams, chairman of the board, had agreed with him after three conferences, that the proposed dam would provide little or no flood control and that it would not provide the amount of power needed to develop the industrial resources of that part of the state.
>
> The Governor emphasized, however, that the board should continue to build new reservoirs in northern Herkimer County to provide better flood control and more power. . . .[17]

Some officials of the District maintain that the two new Board members were instructed to scrap the Higley Mountain proposal and go ahead with the Panther Mountain project. They also assert that certain prominent conservationists agreed not to oppose the Panther Mountain proposal if Higley Mountain plans were dropped, and they are bitter over the subsequent violation of what they regarded as a "gentlemen's agreement."

Whatever the intimate details, the net result was clear: the Higley Mountain reservoir was in limbo. Sides had been chosen, and the adversaries had squared off for the real test. The stage was set for the

[16] *New York Times,* January 31, 1947, p. 3.
[17] *New York Times,* November 14, 1947, p. 26.

big show, the hard-fought and protracted battle over the Panther Mountain reservoir.

The Battle of Panther Mountain

The crucial engagement of the war began officially on December 23, 1947, when the Board of the Black River Regulating District resolved to amend the official plan to provide for a Panther Mountain reservoir with a storage capacity of not less than 12 billion cubic feet (277,000 acre feet, nearly 90 billion gallons). Nine billion cubic feet was allocated to river regulation, three billion to flood control. The reservoir would cover 4,124 acres of land, 934 acres of which were owned by the State of New York as part of the Adirondack Forest Preserve; the remaining 3,190 acres were owned or leased by the Adirondack League Club, a private organization. The proposed amendment shortly thereafter was approved by the Water Power and Control Commission, and on January 13, 1948, the Board resolved to draw up preliminary detailed plans, maps, specifications, and estimates for the reservoir. The preliminary plans received the unanimous approval of the Water Power and Control Commisison on May 4. The Board then proceeded to announce a public hearing on the proposal.

The hearing, held in Watertown in seven sessions between June 8 and July 9, 1948, was of special importance for two reasons. In the first place, it afforded the observer an opportunity to see what individuals and groups lined up for and against the reservoir. The following 36 groups were represented or listed as against the proposal:

Adirondack League Club
Adirondack Moose River Association
Adirondack Mountain Club
American Nature Society
American Planning and Civic Association
Association for the Protection of the Adirondacks
Black River Flats Owners Protective Association
Camp Fire Club of America
City College Hiking Club
11th Ward Fish and Game Club (Utica)
Emergency Conservation Council
Federated Sportsmen Club of Oneida County
Fin, Fur, Feather Club (Utica)
Fish and Game Unlimited (Syracuse)
Forest Preserve Association of New York State
Greenfield Fish and Game Club (Utica)

Guan-Ho-Ha Fish and Game Club (Scotia)
Herkimer County Sportsmen's Unit
National Park Association
National Wildlife Association
New York State Association of the Garden Clubs of America
New York State Conservation Council
New York Division of the Izaak Walton League
New York Zoological Society
Nine Mile Rod and Gun Club (Rome)
Oneida County Forest Preserve Council
Piseco Fish and Game Association
Rome Fish and Game Association
Saratoga County Fish and Game Council
Schenectady Bird Club
The Adirondack Sportsmen's Federation
The Wilderness Society
Trail Hound Association of Rome
Travelers Club of Deansboro
Trenton Fish and Game Association
Utica Radiator Corporation

In addition the opponents numbered several prominent conservationist spokesmen who were in attendance.

The units and organizations represented or listed as in favor of the proposal were:

Carthage Paper Makers
City of Watertown
Lewis County Board of Supervisors
Lewis County Chamber of Commerce
Lewis County Farm Bureau
Town of Denmark
Village of Black River
Village of Brownville[18]

The small number of proponents from the Black River basin in comparison to the large number of non-area opponents was portentous.

The second reason for the hearing's importance was that the arguments for and against the reservoir began to crystallize there. As the hearing progressed the opponents took the line that they were not

<hr>

[18] Supreme Court of the State of New York, Appellate Division, Fourth Department, *Record in Proceedings in the Matter of . . . Adirondack League Club v. The Board of the Black River Regulating District . . .* , Vols. I, II; pp. 143-1024. Hereafter cited as *Proceedings*. Note again should be taken of the preponderance of representation from the Mohawk-Upper Hudson Region.

against a reservoir, but only the Panther Mountain reservoir. They thus attempted to develop the argument that alternative sites were as good as or better than Panther Mountain. Edmond H. Richard, president of the Adirondack Moose River Committee, testified that

> Little or no consideration has been given by the Black River Regulating Board to the merits of other locations. . . . A reservoir at Hawkinsville or one built near Port Leyden would provide better flood control on the Black River than any reservoir to be constructed at the location of the proposed Panther Mountain Reservoir. . . . The . . . Board . . . persists in pressing its Panther Mountain reservoir project without regard to the other more available sites arbitrarily despite the facts. . . .

Richard also echoed the other major arguments of the opponents when he maintained that the reservoir would destroy both the wilderness character of the site and an area essential to wildlife. Another argument that was to be advanced more vigorously at a later period was that the reservoir would take property unconstitutionally under the guise of public use for what was in reality private gain.

The main argument of the proponents was that

> . . . the Panther Mountain Reservoir is the most economical reservoir in the Moose River other than the Higley Mountain Reservoir. . . . Now the Panther Mountain Reservoir is the cheapest and most feasible, measured in terms of benefits.

In addition they argued that other benefits, among them flood control, recreation, and pollution prevention and abatement, would accrue to the inhabitants of both the valley and the rest of the State. Although the hearing was conducted in an atmosphere of informality, the transcript conveys a tenseness, a forced restraint that was soon to break loose. Clearly the situation had reached the point of no compromise.[19]

On November 11, 1948, the District's Board resolved to proceed with the construction of the Panther Mountain reservoir. The decision was based upon twenty-six findings of fact which provide an excellent summary of the Board's position and the factors underlying it.[20]

Almost immediately the reservoir's opponents swung into action to avert construction. Utilizing the provisions of Section 454 of the Conservation Law, the Adirondack League Club and other interested parties sought review of the Board's decision by the Supreme Court of the State of New York. On December 27, 1948, the cases of *Adirondack*

[19] The above testimony is quoted and summarized from *ibid.*, pp. 194-233.
[20] *Minutes*, November 11, 1948.

League Club v. *The Board of the Black River Regulating District* and *Adirondack Moose River Committee, Inc., Edmond H. Richard et al.,* v. *The Board of the Black River Regulating District* were sent to the Appellate Division. The petitioners included most of the groups listed in opposition at the public hearing, reinforced by the Monroe County Conservation Council, Genesee Conservation Council, Oriskany Rod and Gun Club, and five individuals.[21]

There were numerous complaints presented against the District and the action it proposed to take, but stripped of legal verbiage, they were, in brief, that:

1. Article VII of the Conservation Law is unconstitutional because it grants river regulating districts arbitrary and unlimited power to appropriate property in violation of the due process clauses of the United States and the New York State Constitutions;

2. taking property for private purposes is in violation of the laws and Constitution of the State of New York, and the primary purpose of the reservoir, power development, is private;

3. the Board's determination was not made in a civil action or special proceeding by a court of record or a judge of a court of record; and

4. some members of the Board were disqualified to sit because of a personal interest in the matter.

The plaintiffs also contended that the determination had been wrongfully arrived at, production of power does not warrant destruction of the Forest Preserve, maintenance of forest cover is better for flood prevention than reservoirs, periodic floods rejuvenate farmlands, the construction of the reservoir was not in the public interest, and wildlife would be deprived of natural feeding and breeding grounds. They argued further that the New York Constitution required regulating reservoirs to be operated by the State and the river regulating district was not "the State."

In spite of the barrage of charges, the determination of the Board to build the reservoir was upheld.[22] Not one of the conservationists' arguments was sustained. Nevertheless, the District was not free to go ahead with construction because of events in another sector.

On March 23, 1949, the Legislature created a Joint Legislative Committee on River Regulation ". . . to make a thorough study and investigation of the problem of existing and future river regulation within the state with a view to recommending appropriate legislation

---

[21] The summary which follows is based on *Proceedings*, as cited above.

[22] *Adirondack League Club* v. *Board of the Black River Regulating District*, 275 A.D. 618 (1949).

or constitutional amendments. . . ." [23] As it turned out, the meetings held by the Committee had as their chief result the publicizing of the views of the reservoir's opponents. River regulating districts in general were subjected to severe criticism. A major point was reflected in a statement by Herman Forster, President of the New York State Conservation Council, before a meeting of the Committee on September 17, 1949. Forster assailed the autonomy of regulating boards, which he charged operated in the name of the State without having to answer to either electorate or executive branch, except on engineering problems. Another recurrent criticism was voiced at the same meeting by Edmond Richard, of the Adirondack Moose River Committee, who placed his organization in opposition to the "unnecessary and unwarranted flooding of the forest preserve." The committee hearings continued to be anti-Panther Mountain in complexion. It was not that the proponents were denied the right to appear, but simply that the sportsmen's groups through sheer weight of numbers dominated the proceedings. As the hearings continued the anti-reservoir forces picked up even more support. U. S. Senator Herbert H. Lehman and representatives of the U. S. Fish and Wildlife Service, Federation of New York State Bird Clubs, and New York State Federation of Women's Clubs at one time or another voiced their opposition before the Committee. There was no doubt as to the character of the testimony developed by the Committee: the bulk of it was distinctly hostile to the river regulating district.

Nor were the sportsmen and the conservationist groups without energetic friends in the Legislature. State Senator Walter W. Stokes, a member of the Adirondack League Club, introduced a bill to amend Section 445 of the Conservation Law to the effect that "No reservoirs for the regulation of the flow of streams or for any other purpose except for municipal water supply shall be hereafter constructed in Hamilton or Herkimer counties on the south branch of the Moose river by any river regulation board." The bill was passed by the Legislature and sent to Governor Dewey who, under unusually strong pressure from sportsmen's groups and upstate Republican leaders, signed it. [24] The measure became famous as the "Stokes Act." Four days later, the Appellate Division's decision upholding the Board's determination to build the Panther Mountain reservoir was reversed by the Court of Appeals, which cited the Stokes Act in holding moot the questions presented by the case. The Court at the same time refused to rule on the consti-

---

[23] Legislative Document No. 70, 1951, p. 24. The committee was continued from year to year until 1952.

[24] Chap. 803, Laws of 1950. *New York Times,* April 22, 1950, p. 50.

tutionality of the act, declaring it was ". . . a problem which we postpone for later consideration since it is not now before us." [25]

The passage of the Stokes Act brought to a head a problem confronting the Black River Regulating District. In carrying out the preliminary work on the proposed Moose River projects, the district, under existing provisions of the Conservation Law, had accumulated a floating debt of $203,000 in the form of certificates of indebtedness. The certificates were to be redeemed by assessments against the beneficiaries of the Moose River reservoir(s), a source of revenue which appeared more and more uncertain. The Black River Board took the position that the contracting parties had acted in good faith; that the debt had been legally incurred; and that, if the rules of the game were going to be changed in such a way as to deprive the District of its only means of repayment, it was entitled to relief. Three alternative solutions to the floating debt problem were discussed in a conference the Board had with the State Comptroller in his office on November 29, 1950.

They were:

1. impose an apportionment of the costs on the beneficiaries of the proposed reservoir;
2. seek an appropriation from the Legislature; or
3. work for the repeal of the Stokes Act and the construction of the Panther Mountain reservoir.[26]

The last alternative received local support when the Watertown Chamber of Commerce called a conference to examine the possibilities of activating a group to work for Black River regulation and to attack the constitutionality of the Stokes Act.[27]

The next move was taken by the anti-reservoir forces when, early in 1951, Senator Stokes and Assemblyman Justin C. Morgan introduced a proposed constitutional amendment which would require public approval by statewide referendum of all stream-flow projects. The proposal would have shifted the power to decide on such projects from the Legislature to the people. The resolution failed of passage, but it

[25] *Adirondack League Club* v. *The Board of the Black River Regulating District*, 301 N.Y. 219 (1950).

[26] *Minutes* (inclusion, no page number).

[27] Watertown's stake in the Black River and its regulation is a vital one, for it depends upon the river for water supply, hydro-electric power (the city operates a plant from which it produces power to light its streets and public buildings), and sewage disposal.

contained an idea to which the forever-wild forces were to return later.

On May 18, 1951, the District's Board undertook positive action by passing a resolution to go ahead with proceedings for construction of the Panther Mountain reservoir. This move was followed at the end of June by a condemnation action against the Adirondack League Club, and that step in turn by a request for a New York State Supreme Court order declaring the Stokes Act unconstitutional.

The situation which had evolved was unusual. A state agency against whose activities a legislative act had been directed was seeking to void the act. The State on its part maintained that the Board of the District had no right as a creature of the Legislature to challenge the constitutionality of a legislative act. An Assistant Attorney-General, E. L. Ryan, argued that the Court of Appeals decision in 1950 had effectively disposed of the Board's case.

At the beginning of 1952 the locus of the Panther Mountain controversy switched back to the Legislature. There Assemblyman John C. Ostrander, who had served as chairman of the late Joint Legislative Committee on River Regulation, introduced a proposed constitutional amendment to prohibit the construction of river regulating reservoirs in the Forest Preserve. The proposal was, in effect, to reverse a policy adopted by constitutional amendment in 1913. The joint hearing of the Senate and Assembly Conservation Committees on the subject revealed an interesting development; for as the battleground expanded to encompass the whole State, Panther Mountain gained substantial support from non-District quarters. The State Grange, for example, went on record as opposing the Ostrander Amendment, as it came to be known. E. Payson Smith, public relations director of the Grange, charged that the amendment was being pushed by ". . . a small but powerful private ownership group," and maintained that the public interest would be endangered by restricting the means of further river regulation. Smith was joined in opposition to the proposed amendment by representatives of the Empire State Association of Commerce and the Associated Industries of New York State.

Notwithstanding this impressive support, the opponents of the amendment were unable to prevent its overwhelming passage. On March 5, 1952, the Assembly passed the resolution by a vote of 131 to 14. One week later it passed the Senate, 48 to 6. The proposal then had to be resubmitted to the 1953 Legislature for second passage before being presented to the voters for their decision; meanwhile the pro-reservoir forces received another setback from the courts. On March 24, the New York Supreme Court denied the Black River Regulating

District the right to seek an advisory opinion on the constitutionality of the Stokes Act.[28]

By this time the Board had decided to attack from another angle. In February it had hired the public relations firm of Woodard and Voss to ". . . eliminate misunderstanding concerning the purposes of the establishment of the Black River Regulating District."[29] Further, Alleyn H. Beamish was retained as the Board's public relations counsel in June. At this stage of the controversy the District was continually on the defensive. The many sportsmen, conservationist, and garden clubs constituted an effective network of grass-roots organizations. The substance of the argument against the construction of a reservoir within the Forest Preserve was easier to grasp than the more technical pro-reservoir argument. While spokesmen for the Ostrander Amendment could appeal to the naked interests of the clubs' members, its opponents were hard pressed to make complex facts simple. Even so, the proponents of the reservoir continued to gain support. The New York State Federation of Labor backed the Panther Mountain proposal. The State C.I.O. also supported the project, including a statement in opposition to the Ostrander Amendment in its 1953 legislative program.[30]

The pace of charge and counter-charge increased as the 1953 Legislature convened. In the Assembly, the Democratic minority leader promised the support of his party for the amendment. Support was also pledged by Oswald D. Heck, Speaker of the Assembly; Lieutenant-Governor Frank C. Moore; and Francis J. Mahoney, Senate Democratic leader. On the other hand, Governor Dewey expressed grave doubts about the amendment. Nevertheless, the proposal passed by a 122 to 29 vote in the Assembly and a 42 to 11 vote in the Senate.[31] The vote was regarded by some as a major setback for the Governor.

---

[28] *Board of the Black River Regulating District* v. *Adirondack League Club, et al.*, 115 N.Y.S. 2nd (1952).

[29] *Minutes*, February 18, 1952.

[30] *Albany Knickerbocker News*, January 15, 1953. The support of the state body was offset somewhat by what appeared to be considerable rank-and-file opposition to the reservoir. For example, the Greater Utica Industrial Union Council (CIO) passed a resolution charging that the dam was a scheme of the state power trust that would flood the people's land and destroy their natural assets. *Utica Press*, February 4, 1953.

[31] *Albany Knickerbocker News*, March 22, 1953. There were some attempts to link Secretary of State John Foster Dulles to Governor Dewey's stand on the issue. The Secretary, who spent his youth in Watertown, had remained silent on the issue in a special election campaign for the U.S. Senate. *Massena Observer*, April 6, 1953.

While the political clouds were lowering over Panther Mountain in the Legislature, the District's legal counsel continued to seek relief through the courts. The Appellate Division of the Supreme Court again was asked for a declaratory judgment concerning the constitutionality of the Stokes Act. In May the Court reversed the Supreme Court ruling that the Board could not test the act's validity.[32] The Court did not, however, declare the act unconstitutional.

The campaign for popular adoption of the amendment began in earnest once the measure had passed the Legislature. The leader of the opposition forces was Governor Dewey, who was joined by the several organizations listed previously as favorably disposed toward the reservoir proposal, the State Bar Association, State Conference of Mayors, State Public Recreation Society, many city councils, and some twenty-five newspapers (of which only one was published in New York City). The officers of the Black River Regulating District did what they could to defeat the proposition. The District's public relations counsel earlier had created a Citizens Council on Water Conservation as a pro-reservoir organization; and the Board authorized an expenditure of $10,000 for public education, an extremely modest sum in view of the scope and nature of the developing campaign.

Among the more prominent of those who campaigned for the amendment were Paul Schaefer of the Friends of the Forest Preserve, Inc.; Edmond H. Richard, President of the Mohawk-Hudson Federation of Conservation Councils; Herman Forster, President of the Adirondack Moose River Committee, Inc.; Ralph A. Atwater, Chairman of the Oneida County Forest Preserve Council, Inc.; and Lithgow Osborne, President of the Association for the Protection of the Adirondacks. Osborne had been Conservation Commissioner when the Higley Mountain proposal had been approved, a fact his opponents did not ignore. A Council of Conservationists for Amendment 9 made its appearance as a holding group representing sixteen organizations. The aim of the Council was ". . . to coordinate the efforts of 100,000 organized conservationists and sportsmen . . ." in a drive for passage of the Ostrander Amendment.[33] In addition, the proposed amendment re-

[32] *Board of the Black River Regulating District* v. *Adirondack League Club*, 282 A.D. 161 (1953).

[33] *Auburn Citizen Advertiser*, September 24, 1953. There was some disagreement about the total number of conservationists and sportsmen involved. One of the groups, the Adirondack Moose River Committee, had claimed 1,000,000 members, while the National Wildlife Federation had claimed 140,000. See *Proceedings*, p. 31.

ceived support from about thirty-one papers and nearly every sports editor in the State.

On November 3, 1953, the voters were asked:

> Shall the proposed amendment of Article Fourteen, Section Two, of the Constitution, to prohibit the use of portions of the Forest Preserve for the construction of reservoirs to regulate the flow of streams be approved?

The vote in favor of the amendment was overwhelming: 943,200 voted "yes" and 593,696 voted "no." [34]

Three days later the chief engineer of the Black River Regulating District reminded the Board that the Panther Mountain project remained a practical one from engineering and economic standpoints. The Board thereupon concluded to proceed with plans for the reservoir, although it was not until February that a firm decision was reached to seek the adoption of a constitutional amendment providing for a Panther Mountain reservoir. [35] The Board drew up a bill to that effect for introduction in the Legislature and retained a legislative representative to handle the "selling" aspects. The proposed amendment came to be known as the Wise-McGuiness bill, from its sponsorship by Senator Henry Wise and Assemblyman James J. McGuiness. It sought 1,500 acres of Forest Preserve land for a reservoir on the south branch of the Moose.

The proposal had rough sledding in the Legislature. At the public hearing on conservation bills, Lithgow Osborne leveled a broadside of criticism against measures that would permit construction of highways and utility lines, the Panther Mountain reservoir, and a state park near Old Forge, all of which would involve Forest Preserve lands. He stoutly defended the "forever wild" principle embodied in the Constitution. Charles H. Tuttle, legal counsel for the Adirondack League Club, expressed open indignation, noting that:

> For nearly four years the Black River Regulating Board, consisting of three local administrative officers (appointed but not elected), has been defying the entire state government, the Legislature (which adopted the Stokes Act forbidding the dam), the governor who signed the Stokes Act . . . and the attorney general, who both in and out of court has been upholding the constitutionality of the Stokes Act. [36]

The conservationists also employed the time-honored delaying tactic of

[34] *New York Times*, November 5, 1953, p. 26.
[35] *Minutes*, February 1, 1954. The proposal was introduced in the Legislature on February 15, 1954.
[36] *Rochester Times-Union*, March 9, 1954.

proposing that the Panther Mountain reservoir question be referred to a legislative committee for study.

On March 16 the proposal failed by one to secure the 29 votes needed for passage in the Senate. A strong coalition, however, rallied to its support. Governor Dewey, Daniel P. O'Connell (Albany Democratic leader), private power interests, and up-state legislators joined forces to keep the measure alive. In the lower house, Assemblyman Ostrander argued that the voters had decided the issue the previous November. Others took the view that it was only fair play to let the people decide on the specific question. The latter view prevailed and the bill passed, 117 to 75. The show of strength in the Assembly together with the pressure generated in the Senate forced the latter body to reconsider its stand and pass the bill. The vote was 29 to 21.

In July the Court of Appeals dealt the District a sharp blow. The Court, in answering the argument that the District was deprived of contractual rights by the Stokes Act, said, in part:

1. Although the Stokes Act and Ostrander Amendment were made after the preliminary approval of the reservoir, they nevertheless enunciate public policy;
2. the withdrawal of Forest Preserve land by the Ostrander Amendment was in no way limited by the district's claim to vested right to proceed on a basis of prior approval;
3. the number and nature of the district's powers are within the absolute discretion of the State, and any alteration or impairment of the powers by the Legislature is constitutional and not protected by the due process clause of the 14th Amendment of the U.S. Constitution;
4. even though the district was voluntarily formed, it is an agency of the State;
5. inclusion of the reservoir proposal in the official plan does not create a contractual interest in the site that could not be impaired by legislative action, the interests of the district being the interests of the State and the State does not challenge its own acts;
6. the certificates of indebtedness do not confer an independent status to test the validity of the Stokes Act.[37]

The Court later denied the District permission to re-argue the case against the constitutionality of the Stokes Act.

Presumably the District had exhausted its legal resources, hence the only course left open was that offered by constitutional amendment.

---

[37] *Board of the Black River Regulating District* v. *Adirondack League Club*, 307 N.Y. 475 (1954).

The proposed amendment (the Wise-McGuiness bill) passed the Senate (30 to 28) and the Assembly (103 to 43) for the second time in the 1955 session. This action made it possible to bring the proposal to popular vote in the statewide election of November 8, 1955.

The campaign for and against the amendment was very much like that of 1953. The Board hired a specialist to conduct the battle for the amendment, and authorized expenditure of $17,100 for a public information program. The chief convert to the cause was Senator Herbert Lehman, who had supported the Ostrander Amendment in 1953. Newspaper support dropped slightly. The conservationists continued to fight the Panther Mountain proposal as posing the threat of commercial exploitation of the Forest Preserve. They enjoyed considerably more newspaper support than in the earlier campaign, and to the Panther Mountain partisans appeared to have all the money they needed. A newspaper of the region commented thus on the campaign:

> One thing—somebody with lots of money must be fighting Panther Dam, judging by the countless letters and brochures that flood this office. And lobbies with lots of money make us do no end of wondering.[38]

The outcome of the uneven contest could have been foretold long before the votes were counted; but to make the record complete the popular vote was 613,927 for, 1,622,196 against. The amendment carried only in Jefferson and Lewis, which as downstream Black River counties would have benefited most from construction of the dam.

By way of anti-climax, the District appealed the ruling of the State Court of Appeals to the United States Supreme Court. In May 1956 the Supreme Court dismissed the action ". . . because the judgment [of the state court] rests on an adequate non-federal basis." [39]

Epilogue

If there is a stream in the State that warrants action under the river regulating district law, that stream is the Black River; yet excepting for three modest reservoirs built three-quarters of a century ago (one

---

[38] "Panther Dam Again" (editorial), *Newark Valley Herald*, October 21, 1955. Leaders of the opposition confess that the defeat of the Higley and Panther Mountain proposals cost a good deal of money. Much of it came from individuals, some of whom gave substantial sums, but the bulk was contributed by local sportsmen and conservation clubs. They are proud of the grass-roots strength of their cause.

[39] *Board of the Black River Regulating District* v. *Adirondack League Club*, 351 U.S. 922 (1956).

of them since enlarged), the River remains unregulated. How might regulation be brought about? What alternatives are available to the Black River Regulating District? For one thing, it might explore somewhat more carefully other reservoir sites—its critics charge that it has failed to do this, but instead has placed its faith too largely in Panther Mountain. In defense of its course the District can point out that, in settling upon Panther Mountain, it chose the most economical and the most feasible site available, one moreover that was approved by both the U.S. Army Corps of Engineers and the New England-New York Inter-Agency Committee. Second, it might seek to bring about the repeal of the Stokes Act, which prohibits construction of regulating reservoirs in Hamilton and Herkimer Counties. This would permit construction of a dam at Panther Mountain, although the reservoir could be backed up only two miles to Combs Creek, which marks the Forest Preserve line. This would be reckoned a moderately successful solution of the problem, though from the evidence at hand it would be difficult to achieve. Third, the District might try again to amend the Constitution to permit construction of storage reservoirs in the Forest Preserve. The best it could hope for would be the restoration of the constitutional provision of 1913, and in view of recent events this would seem a forlorn hope. In the circumstances, alternative three appears unattainable and alternative two difficult.

The practical question, then, which confronts the District—and by inescapable inference, any river-regulating district—is this: what reservoir sites are available outside the Forest Preserve? The leaders of the conservationists who opposed Panther Mountain insist that there are several tenable sites. Moreover, they aver that much could be done to take the tops off floods through re-forestation, particularly on the Tug Hill plateau. Their concern, they maintain, is to protect the Forest Preserve, otherwise they say they would like to be of help to the District. The members of the Board and people up and down the valley find this difficult to believe. They profess to find no evidence of a spirit of cooperation in those who beat them so soundly at the Battle of Panther Mountain.

The next move is up to the Board of the Black River Regulating District, which finds itself in a quandary. Its best efforts to date have proved not good enough. It must do something, but the path before it is far from inviting. And while the Board ponders its plight, the River runs its accustomed course unmindful of the receding din of battle. In the spring the floodwaters take over and the farmers in the flats below Lyons Falls move back into the hills. In summer the water wheels grind to a halt, while downstream Watertown combats the effects of low flow with chlorine. In the valley of the Black, as almost nowhere else in New York, the problem is too much water at one season and too little at another. The ratio of high flow to low flow at Watertown has been as high as 34:1 in modern times. Panther Mountain would not have equalized this flow, but it would have helped.

CHAPTER VII

Stream Pollution: The Buffalo River Case

EACH day Lake Erie empties into the Niagara River enough water to supply two-thirds of the country's total need for that day, except for water power. In 1933 a slug of domestic sewage and industrial waste, activated by a combination of storms and water level changes, moved out from Buffalo harbor and into the Niagara. Down the length of the river it flowed, and out into open water and along the south bank of Lake Ontario. There were outbreaks of gastroenteritis and other water-borne sicknesses in Tonawanda and Niagara Falls, and there was a measurable amount of contamination at water intakes well into the St. Lawrence River, hundreds of miles from the source of pollution at Buffalo. The honeymooners who, if the script is correct, stood en-raptured above Horseshoe Falls, looked down on one of the most polluted streams on the North American Continent. The principal villain in the piece was a minor tributary, the Buffalo River, whose un-appetizing history is the subject of the present discussion.

THE NIAGARA FRONTIER

The term "Niagara Frontier" had a fairly precise meaning in earlier days, when it was employed to refer to a tract of land purchased by Sir William Johnson in behalf of the British Crown from the Seneca Indians. Currently it is used to apply generally to the western half of Erie and Niagara Counties. The chief feature of the Frontier nowadays, apart, perhaps, from the Falls, is the metropolitan complex which, centering on Buffalo, swings northward to include the Tonawandas and the City of Niagara Falls. In this area resides a population of more than 1,100,000 (1952 estimate), of which over 80 per cent is urban. The major city is Buffalo, with a population upward of 600,000, but there are half a dozen other cities with populations ranging from 15,000 to 100,000 or more.

If the Niagara Frontier is symbolized by the metropolis, the latter in turn is characterized by its high degree of industrialization. Pure and plentiful water, cheap hydroelectric power, and good transportation facilities have drawn a great variety of industries to the area. Chief among the industrial products are chemicals, paper and fiber, steel, oil, and coke and its by-products. From the days when the Senecas dickered with Sir William, the Niagara Frontier has developed into one of the largest industrial concentrations in the United States. The industry of the area uses tremendous quantities of water—one estimate places the industrial withdrawal of water in the vicinity of a billion and a half gallons a day. More to the point of the present discussion, industry also is a gross polluter of the very water by which it lives.[1]

By any standard except the one relevant to this study, the Buffalo River is not an important stream. Formed by the junction of two small creeks, Buffalo and Cayuga, it is joined by Cazenovia Creek before it empties into Lake Erie at the point of origin of the Niagara River, to which it is regarded as a tributary. Its drainage basin covers about 435 square miles, an area less than half that of the average American county. While the three creeks are pleasant, lively streams, the River itself is little more than a dredged channel extending inland some six miles from Buffalo harbor. Originally 30 feet wide and 7 feet deep, it is now 200 feet wide and 22 feet deep. It is this artificial basin that is important to the present story.

For present purposes, the Buffalo has three outstanding features. First, it has virtually no current during the summer months, when it becomes little more than a stagnant pool. It has experienced daily flows of less than ten cubic feet per second. Second, it suffers an occasional flash flow, due to heavy rains upstream or to a Lake Erie seiche (a strong easterly wind which lowers the surface of the lake), or both, which increases its discharge to as high as 27,000 cubic feet per second. The floods are not especially destructive, but the extreme variation in flow serves first to accumulate great slugs of pollutants in the River, then to purge them into Buffalo harbor and thence into the Niagara. The problem is further complicated by the fact that, for weeks at a time during the winter, the wastes are caught up at the entrance to

[1] Charles W. Reck and Edward T. Simmons, *Water Resources of the Buffalo-Niagara Falls Region* (Geological Survey Circular 173, Washington, 1952). This publication provides a very useful introduction to the Niagara Frontier.

the Niagara in a great ice pack, to be released in one grand deluge of pollution with the spring thaw. Third, there are several major industries located on the river, and they provide the pollutants which consolidate into slugs.[2]

The Buffalo River's reputation as a highly polluted stream is ancient and secure. In the early days of the century, before treatment of domestic sewage, it was in part responsible for the high typhoid rate which prevailed from Buffalo north along the Niagara. Note has been made of the outbreak of gastroenteritis and other water-borne illnesses in 1933. The Department of Conservation has reported fish kills (the sudden death of large numbers of fish without apparent cause) on numerous occasions—on May 25, 27, and 28, 1943, April 15 and 16, 1948, and March 24, 1953, for example. The agent of destruction was not identified in every case, but in the last two cyanide was named as the culprit.

Pollution in the Buffalo River may be said to be of three kinds. First (and least important) is silt, which the State Department of Health has passed by as a pollutant in favor of emphasis on more important sources. Nevertheless the Corps of Engineers, which is responsible for keeping the River navigable, and at least one of the big industries to be mentioned presently have expressed interest in silt reduction. Second, while municipal sewage is not as important a source of pollution as it was three decades ago, it is still one to be reckoned with. Most of Buffalo's city sewers receive both sanitary sewage and run-off water, and during periods of heavy rainfall they are not equal to the added burden placed upon them. Moreover, it is standard practice to discharge heavy overflows into the River with no treatment at all. Hence the problem of pollution from domestic sewage remains, though in modified form.

Third and by all odds most important are industrial wastes, which originate in a major industrial development some five miles inland from the mouth of the Buffalo River. There, in a space of approximately one mile, five great industries are located along the banks of the stream. These are the Socony Mobil Oil Company, the General Chemical Division of the Allied Chemical and Dye Corporation, the National Ani-

---

[2] Hayes H. Black and Earl Devendorf, *Industrial Wastes along the Niagara Frontier and their Effect on the International Boundary Waters* (a paper presented at the Twenty-sixth Annual Meeting of the New York Sewage and Industrial Wastes Association, New York City, January 21-22, 1954), pp. 6-7. As the title indicates, this publication is directly relevant to the problem at hand.

line Division of Allied Chemical and Dye, the Republic Steel Corporation, and the Donner Hanna Coke Corporation. Industrial discharges into the Buffalo River include acids, oils, sulphur compounds, phenols, cyanides, and other chemical wastes. The resulting pollution is largely chemical in character, as distinguished from the bacterial pollution of municipal sewage. Chemical pollution is much less dangerous as a cause of human disease than bacterial pollution, for its threat to health can be largely counteracted by treatment. Some chemicals, however, leave an unpleasant taste even in treated drinking water. Moreover, such treatment is expensive, which is one of the major reasons for the movement to reduce industrial pollution. Another is found in the fact that such pollution detracts from and indeed often destroys the recreational use of an area; for it is offensive to sight and smell, even when not positively harmful. Another still is found in the constant danger to fish life, which endures a precarious existence in chemically polluted waters.

Following the unusually severe pollution of 1933 and the resulting widespread illness, the State Commissioner of Health in 1934 issued an order requiring the City of Buffalo to take measures to abate domestic and industrial pollution in adjacent waters. The result of the negotiations following this order was a legislative act, passed in 1935 (Chap. 349, Laws of 1935), which created the Buffalo Sewer Authority, a public benefit corporation charged with the responsibility of pollution control. The Authority proceeded with its task, and in 1938 placed in operation a $15,000,000 plant for the treatment of domestic and industrial sewage. Immediately thereafter industrial wastes carried by the Buffalo River fell from the equivalent of the sewage of a city of 244,000 persons to that of a city of 100,000. After the war the region experienced a great industrial expansion, with a consequent serious increase in pollution. A study made in 1946 concluded that "Under conditions of low flow [the Buffalo River] becomes grossly polluted and is a vast septic tank in which nuisance conditions are created." The population equivalent of pollution had climbed to 221,000. From that year on the population equivalent gradually decreased, due in part to pollution reduction efforts by the industries themselves, and by 1953 the figure stood at about 175,000. Conditions were not as bad, therefore, as they had been twenty years before, but they were still far from satisfactory. The Buffalo River remained a badly polluted stream.[3]

[3] Erie County Health Department, *Water Pollution in Erie County* (Buffalo, 1953, mimeographed.)

THE PROBLEM OF POLLUTION ABATEMENT

A program of pollution abatement is difficult of achievement in the best of circumstances; the problem is further complicated on the Niagara Frontier because of its international aspects. The Boundary Waters Treaty of 1909 provides (in Article IV) that neither the United States nor Canada may pollute international waters to the injury of health or property on the other side. The International Joint Commission established by the Treaty has kept a watchful eye on pollution trends, and from time to time has lent significant aid to abatement programs.[4]

As a purely domestic issue, the problem of pollution in the Buffalo River has had the attention of no less than twelve governmental agencies. Four of these are federal. The U.S. Public Health Service has interested itself for many years, particularly in connection with the work of the I.J.C. The U.S. Geological Survey conducted an important study to which reference was made earlier (footnote 1). The Bureau of Fisheries made a special study in 1938-39 and issued a report on "Industrial and Organic Pollution of the Buffalo River." The Soil Conservation Service is engaged even now in a water control and land treatment program in the tributary streams.[5]

State agencies concerned with the problem include the Joint Legislative Committee on Natural Resources and the companion Committee on Interstate Cooperation. The interest of the first stems from its preoccupation with water as a natural resource and from its specific concern for water pollution. The second has maintained an active interest in interstate affairs for many years, and so has watched the Buffalo River situation with an eye to possible extra-state involvement. The concern of the State Department of Conservation is for pollution abatement to preserve fish life; it has made a number of studies of this subject in the Erie-Niagara region, and more than once has centered its attention on the Buffalo River. The State Department of Health is charged with primary responsibility for the avoidance and abatement of water pollution. Its part in attempting to resolve the Buffalo River problem will be examined with some care later on.

Local units with an interest in the Buffalo River include first the

---

[4] The composition and activities of the International Joint Commission are discussed briefly in Chapter VIII, below.

[5] This undertaking is projected for completion over a period of 18 years at a cost of $4,000,000. Its principal purpose is to reduce siltation in the Buffalo watershed.

City of Buffalo, which was taxed by the State with responsibility for local pollution in 1934 and which has been directly involved in the cause of pollution abatement since. The Buffalo Sewer Authority, created in 1935 to build and operate sewage treatment facilities in behalf of the City of Buffalo, has been active in seeking a solution to the problem of pollution in the area for over twenty years. Erie County, and more especially the Erie County Health Department, likewise have been actively concerned with the problem since 1948, when the Health Department was formed.

In addition to the official bodies, the industries themselves had a direct and substantial stake in pollution abatement in the Buffalo. Finally, a number of private organizations, among them the Erie County Sportsmen's Alliance, the Niagara Frontier Chapter of the Izaak Walton League, the Associated Industries of New York, and the Buffalo Chamber of Commerce, took a lively interest in the proceedings. One of the most interesting aspects of the Buffalo River pollution case is to be found in the number and variety of organizations, both public and private, which became involved in a problem which on the surface did not appear to be particularly complicated.

Official interest in pollution in the Buffalo area goes back to a 1907 survey conducted by the State Department of Health, which sought to determine the suitability of the Niagara River as a source of water supply for Niagara Falls. A similar survey was conducted six years later in connection with the water supplies of Tonawanda and Lockport. The studies concluded that the Niagara was polluted as far down as Niagara Falls, and named the City of Buffalo as the chief (though not the only) culprit. Two 1928 studies, one made by the Department of Health, the other by the Department of Conservation, found a high degree of pollution in eastern Lake Erie and the Niagara River, and listed Buffalo as the chief source of the domestic sewage and industrial waste discovered there. Yet another study by the Department of Health (in 1931) noted marked increase in pollution in the Buffalo area, citing industrial growth as a principal factor.[6]

It was the near-catastrophe wrought by the wave of pollution of 1933 that finally produced specific action. Note has been made of the establishment in 1935 of the Buffalo Sewer Authority, and of the temporary reduction in pollution which followed completion of the Authority's sewage treatment plant in 1938. With a battle won but a

[6] State Department of Health, Report No. 3, *Lake Erie-Niagara River Drainage Basins: Recommended Classifications and Assignment of Standards of Quality and Purity For Designated Waters of New York State* (Albany, 1953), pp. 10-12.

war developing which might yet be lost, the Sewer Authority in 1946 engaged George E. Symons to make a study of pollution in the Buffalo River. The resulting report proved important for both its findings and its recommendations.[7]

The Symons report began with an examination of the conditions of pollution, which it found not so bad as in 1935 but much worse than in the years immediately following 1938. The industries had taken some steps in the direction of pollution abatement, the report continued, but what they had done had not been nearly enough. What was required now was a frontal attack on the problem. The program outlined contained the following major proposals:

1. Adoption of a plan to bring water from Lake Erie for use for cooling purposes by the five major industries located on the Buffalo River.

2. Active participation by the five industries in the Lake Erie water project.

3. Acceptance of the proposition that, while the Buffalo River will never again be a trout stream, pollution nevertheless must be held to a reasonable level. (The level was spelled out in terms of scientific standards.)

4. Acceptance by the industries of the cost of waste treatment as a legitimate part of production costs.

5. Adoption by the industries of the means available to reduce their biochemical oxygen demand (B.O.D.: a measure of the consumption of oxygen in the water) loads by 50 per cent, to eliminate toxic substances where possible, and to "practically completely eliminate the discharge of iron" into the River.

6. Adherence by the industries to the allowable standards of pollution set for the River.

7. Construction by the Sewer Authority of certain sewers to facilitate acceptance of "such wastes from these several industries as may be deemed advisable to be taken into the city sewers."[8]

The heart of the program, the report went on to say, was the plan for bringing cooling water from Lake Erie to the five major industries on the River. The proposal was to bring not less than 120 million gallons of water a day from the lake to the industries, which, having used it for cooling purposes, would then discharge it into the Buffalo River.

---

[7] Buffalo Sewer Authority, *Final Report on Buffalo River Stream Pollution Studies in 1946* (Buffalo, 1946, mimeographed). This is usually referred to as the "Symons Report." It is hereafter so cited.

[8] *Ibid.*, pp. 12-13.

This would create a more uniform water flow in the River, and so would reduce the threat from the slugs of wastes which accumulate during the summer low flow. It would also substantially improve the oxygen content of the water, and would dilute the wastes emptied into the river. From the point of view of industry, it would provide an adequate amount of excellent cooling water, which in turn would reduce maintenance and operating costs. The plan thus would benefit both the industries and the public. Although properly regarded as only one aspect (albeit an important one) of a total pollution abatement program, the cooling water project came to be considered by some to be not only the heart, as Symons had termed it, but the whole of the program.[9]

The Sewer Authority followed up this report with a series of conferences with the five industries looking to its implementation. The Authority's consultant, George Symons, was its representative in these negotiations. Some 18 months after the original report, he was able to make a supplementary report on progress. The industries, he noted, had accepted the proposed program in principle, and specifically had agreed that the cooling water project was feasible and desirable. A number of questions remained, among them that of financing the project, but Symons expressed confidence that satisfactory answers would be found. While the search for an ultimate solution continued, he noted with satisfaction, the industries were taking steps to reduce their discharge of toxic wastes along the lines he had suggested.[10]

Meanwhile, the International Joint Commission, on request of the governments of Canada and the United States, had undertaken a study of the pollution of boundary waters in the Lake Erie-Niagara River-Lake Ontario section. To direct the study, the Commission set up a Board of Technical Advisors consisting of four sanitary experts, two state (or provincial) and two Federal, from each country. The technical work on this side was done by the United States Public Health Service under the direction of Hayes H. Black, Public Health Service engineer and member of the Technical Advisory Board. Mr. Black immediately established contact in behalf of the Board with the interested state and local agencies, and with industry as well. The studies of the technical staff were supplemented by public hearings held in Buffalo on November 15 and 16 and December 13-15, 1949, at which the

[9] *Ibid.*, pp. 14-15.

[10] "Progress since Final Report on Buffalo River Pollution Studies," a mimeographed memorandum to the Buffalo Sewer Authority from George E. Symons, Consultant, June 27, 1948.

Buffalo Sewer Authority's abatement program for the Buffalo River was reviewed at length. Again the five industries accepted in principle the reduction by them of toxic discharges into the River and the cooling water project. With regard to the latter, the hearings pointed up several problems which will be analyzed in another connection. Concerning the problem of financing, the Sewer Authority sought liberalization of the National Water Pollution Control Act (Public Law 845, 80th Congress, 1948) to allow Federal participation in financing a project of this size; and in January of 1950 Representative Tauriello and Senator Lehman introduced bills to that end.

The study conducted by the Board resulted in a report in 1951 which was adopted by both governments. A prominent feature of the report was a set of "Objectives for Boundary Waters Quality Control." The enterprise left a permanent residue in the form of an Advisory Board, appointed by the International Joint Commission, to examine the problem of industrial wastes, with special attention to efforts to meet the standards adopted for international observance. The Board conducts inspections and reports its findings twice yearly to the Commission.[11]

That the interest of the International Joint Commission was to be active and continuing was indicated by the fact that, in April of 1952, the Buffalo Sewer Authority received the I.J.C. report mentioned above, and along with it a letter requesting information on pollution abatement work under way. The B.S.A. reply gave detailed information on the work being done by industry, and concluded with this statement:

> While the Buffalo Sewer Authority has sponsored the Pollution Abatement Project with the pumping of Lake Erie water to the five industries on the Buffalo River, we legally are prevented from the actual construction of this station. We have, however, endeavored to interest some other agency to undertake the project. At the present time negotiations are pending with the Erie County Water Authority, a newly formed agency, which may be equipped to undertake construction and operation of the water pumping system.[12]

Meanwhile, interest in the problem on this side remained at a high pitch, with a number of public agencies active in the field and with the industries themselves continuing their individual abatement efforts.

[11] Black and Devendorf, *op. cit.,* pp. 1-3. This reference is to a summary of the above developments. The Buffalo Sewer Authority was good enough to make the minutes of its Board meetings available to the writer. Statements made about the Sewer Authority rest on these minutes except where otherwise documented.

[12] Buffalo Sewer Authority, *Minutes,* p. 17522.

An Erie County Health Department sanitary engineer, calling attention to a recent serious fire on the River due to oil leakage from the Socony Mobil plant, filed a vigorous report on the subject with the U.S. Public Health Service. The Corps of Engineers undertook a study of iron flue dust in the River in the summer of 1952, but discontinued its investigation some months later pending the outcome of a court case on the subject filed in Chicago.

In the late spring of 1952 arrangements were made for an exploratory conference on the Buffalo River pollution abatement and water supply project. The Conference, which was sponsored by the Joint Legislative Committee on Natural Resources and the Joint Legislative Committee on Interstate Cooperation, met on September 19. It was well attended, as meetings sponsored by one or both of these committees usually are. The conference was noteworthy for several reasons. First, it indicated the continuing interest of these two important committees in the Buffalo River pollution problem. Second, it focused attention, official, industrial, and public, sharply on the problem. Third, it pushed community thinking a little closer to the point of decision. Fourth, it indicated anew the variety of organizations and agencies interested in the Buffalo River. No fewer than a dozen individuals participated actively in the discussion. They represented the Board of Technical Advisors of the International Joint Commission, the State Department of Health, the State Department of Conservation, the Water Pollution Control Board, the Water Power and Control Commission, the State Department of Audit and Control, the Buffalo Sewer Authority, the City of Buffalo, the Erie County Health Department, the Corps of Engineers, Associated Industries of New York, and, of course, the five major industries involved.

The spokesman for the industries, C. C. Coakley of the National Aniline Division of Allied Chemical and Dye, presented the views of industry in a forthright statement. Coakley addressed his remarks principally to the Lake Erie water project, which he regarded as primarily a pollution abatement scheme. As a secondary purpose, he continued, "The industries have accepted the idea that instead of making such a project a straight expense, it might be possible for them to use this water to some economic advantage, before it is discharged into the river." He raised the issues of sponsorship, and more specifically, financing, offering the opinion that the Buffalo Sewer Authority was the logical agency to build and operate "this project of pollution abatement." As a concluding point, he drew attention to the need for "better and more up-to-date figures," noted that the

Sewer Authority had stated that a qualified engineering firm had offered to make a survey and report for $15,000, and pledged industry's cooperation in the event a fresh study should be thought desirable.

The September 19 conference produced the first clearly articulated statement of an issue which had grown increasingly sharp since the Symons report of 1946. The issue, simply put, was this: was the primary purpose of the Lake Erie water project pollution abatement or industrial water supply? If the first, one line of action, one focus of responsibility, one basis for arranging financing would be indicated; if the second, quite a different line of attack might seem appropriate. One of its sponsors reported the results of the conference discussions in these words:

> At the conference, it was the consensus of opinion that, if such a project were to be carried out it would have as its primary function the improvement of pollution conditions in the Buffalo River and as its secondary purpose the supplying of clean, useful water for the industrial operations of the five manufacturing firms.[13]

Discussion of industry's viewpoint as presented by Mr. Coakley produced general agreement on the proposition that a further engineering study of the Buffalo River pollution problem would be useful. A representative of the City of Buffalo made the point that, since the industries created the pollution, they ought to be willing to bear the cost of the survey. The industries accepted the challenge on the spot, agreeing to finance the study. In the weeks following arrangements were completed, and on January 5, 1953, a contract was signed with the engineering firm of Nussbaumer, Clarke, and Velzy, of Buffalo. The firm was directed to re-examine the Symons report in the light of changed conditions, and more particularly to study and report on the problems of design and cost of the Lake Erie water project.

Meanwhile, the Joint Legislative Committee on Natural Resources, accepting the judgment that the Buffalo Sewer Authority was the agency best suited to execute the project, drafted a bill to provide for the necessary extension of the Authority's legal power. The draft was published and circulated among interested parties for comment. An unexpected development, however, caused the Committee to hold

---

[13] Joint Legislative Committee on Natural Resources, *New York State's Natural Resources* (Legislative Document, 1954, No. 72), p. 145. The conference was reported at length in the Committee's 1953 report (Legislative Document, 1953, No. 69), at pp. 79-82; in the 1954 report at pp. 144-146.

its bill up for further consideration. The development took the form of a question whether the Buffalo Sewer Authority was, after all, the agency which should undertake the project. Perhaps it might be handled more expeditiously by the City of Buffalo, it was suggested. Further, the industries demurred on yet another count, suggesting the possibility that the fresh water to be brought from Lake Erie might be pumped into the Buffalo River well above the plants, which might then obtain improved water from the River at considerably less cost than if industry should undertake the expense of the development as a cooling water project. These continued discussions of old issues convinced the Joint Committee that local interests had not yet reached the point of agreement on a course of action, and it concluded to suspend action until local differences should have been resolved.

An Approach to a Solution

It is necessary now to introduce into the story a new and important state agency. In another place (Chapter II, Water Quality section) we have traced the development of a state policy on water pollution control, and have seen how the Water Pollution Control Board came into being. The Board, established by law in 1949, was in operation as a going concern by the first of the following year. Proceeding with caution and feeling its way gingerly among long-established entities, the new agency soon established itself as a primary factor in the Buffalo River negotiations. In turning attention to the Water Pollution Control Board, it is well to begin with an examination of its duties and method of operation.

The Water Pollution Control Law of 1949 provides that

> . . . The Board after proper study, and after conducting public hearings upon due notice, shall group the designated waters of the state into classes. Such classification shall be made in accordance with considerations of best usage in the interest of the public and with regard to the considerations mentioned in Subdivision 3 hereof. (Public Health Law, Section 1209, Subdivision 2.)

Subdivision 3 lists such factors as the physical features of the water to be classified, the character of the surrounding neighborhood, and the uses to which the water is currently being put.

The law also provides that the Board shall have the power to ". . . make, modify, or cancel orders requiring the discontinuance of the discharge of sewage, industrial wastes or other wastes into any waters of the state . . . ," in harmony with the classification adopted. It invests the Board with legal sanctions, providing that it may institute proceedings in a court of competent jurisdiction to compel compliance with its rulings. Recognizing the value of the gentle touch, the law at the same time admonishes the Board to encourage voluntary cooperation in preventing and abating pollution, to abet cooperative action to that end, and to cooperate with other public agencies interested in the field.

Proceeding to equip itself for the discharge of its responsibilities, the Water Pollution Control Board on October 25, 1950, adopted a set of classifications and standards for the governance of water quality.[14] The classes adopted for fresh surface waters, with indicated best usage, follow:

AA: Source of water supply for drinking, culinary, or food processing purposes, and any other usages.

A: The same, with treatment as required.

B: Bathing and any other usages except those set forth in AA and A.

C: Fishing and any other usages except those set forth in AA, A, and B.

D: "Agricultural or source of industrial cooling or process water supply and any other usage except for fishing, bathing, or as a source of water supply for drinking, culinary or food processing purposes."

E: "Sewage or industrial wastes or other wastes disposal and transportation or any other usages except agricultural, source of industrial cooling or process water supply, fishing, bathing, or source of water supply for drinking, culinary or food processing purposes."

F: Sewage or industrial wastes or other wastes disposal. Classes D and E are described verbatim because they are the two concerning which controversy arose later.

In the prosecution of its duties the Board has adopted a procedure which incorporates six major steps. They are:

Step 1: A survey of the drainage basin involved to develop data on which to base sound action.

Step 2: Preparation of a survey report to serve as a basis for the required public hearing (the report must contain the proposed classification plan for the waters of the basin).

[14] Department of Health, Water Pollution Control Board, *Rules and Classification and Standards of Quality and Purity for Waters of New York State* (October 25, 1950).

Step 3: A public hearing on the proposed classification plan.

Step 4: Official adoption of the classification plan, as modified in the light of facts developed by the public hearing.

Step 5: Development of a comprehensive plan for the abatement of pollution in the waters of the basin. The plan sets forth in detail the pollution problem confronted by each municipality, industry, or other entity and specifies the corrective steps to be taken by each.

Step 6: Enforcement of the plan. Here the Board seeks the voluntary cooperation of every organization or unit involved, resorting to use of the legal sanctional power which it possesses only as a last resort.[15]

In 1952 the Water Pollution Control Board was still gathering momentum and had not yet reached full stride. Moreover, it had committed virtually its full resources to a survey of a section of the Hudson River, and so was quite preoccupied. The Buffalo River problem clearly required attention, however, and the Board found a way to launch a survey (Step 1, above) even in the face of heavy prior commitments. This it did by utilizing the staff and facilities of a number of other agencies. Accepting an offer of assistance by the Advisory Board to the International Joint Commission, it launched its survey of the Lake Erie-Niagara River Drainage Basins in the spring of 1952. The survey report was made public approximately one year later. The letter of transmittal accompanying it (dated May, 1953) contains the following account of inter-agency cooperation:

> Engineering and laboratory personnel were provided by the U.S. Public Health Service under the direction of Engineer Richard A. Vanderhoof. These personnel, utilizing the laboratory facilities of the Buffalo Sewer Authority, did the greater part of the field work in connection with this survey. Active assistance was given by engineering personnel of the Erie County Health Department and the Buffalo Regional and Batavia District Offices of the State Health Department. The entire field work, including engineering and laboratory work, related to Eighteenmile Creek Drainage Basin was done by the Erie County Health Department. Special studies in relation to fish and fish culture were coordinated with the other work and conducted by personnel of the New York State Conservation Department.[16]

[15] The Joint Legislative Committee on Natural Resources set forth these procedural steps in detail in its 1954 report, *New York State's Natural Resources* (cited above), pp. 149-150.

[16] State Department of Health, Water Pollution Control Board Report No. 3, *Lake Erie (East End)-Niagara River Drainage Basins: Recommended Classifications and Assignment of Standards of Quality and Purity for Designated Waters of New York State* (Albany, 1953) p. 5.

The description of the Buffalo River from the five industries to its mouth was short and pointed. It read: character of district, industrial; condition of waters, grossly polluted; present usage, industrial water supply and navigation; best usage, industrial water supply; class, D.

A public hearing on the proposed classification plan was called for Buffalo for October 21, 1953. There three major points of view were presented. First, a representative of the Water Pollution Control Board explained why the Board had proposed a Class D rating, and defended that classification. Second, spokesmen for several sportsmen's clubs complained that the proposed Class D was too low and argued for a higher classification: they wanted water fish could live in and they could fish in. Third, spokesmen for the Buffalo Chamber of Commerce and for the industries themselves protested the Class D rating, calling instead for Class E. The National Aniline Division of Allied Chemical and Dye protested (through a brief) that it had already spent $2,000,000 "solely for the benefit of the community." It stated further that installation of the treatment works necessary to conform to Class D standards would cost the company $5,500,000, plus $1,000,-000 annually for operation, and all for the maintenance of the Buffalo River as a fishing stream, ". . . which is no longer an appropriate use of these waters." Associated Industries of New York, noting that the industries had spent $3,000,000 for pollution reduction, stated that they could not continue in operation if the Class D rating should stand. C. C. Coakley, speaking for the five industries, read a long joint statement which concluded that Class D ". . . is economically impossible of attainment by the Buffalo River industries." A separate report by a consulting sanitary engineer, filed at the instance of the industries, supported their case. The engineer concluded that a D classification would cost the industries $7,000,000 in money, apart from the cost of acquiring the 25 acres which would be needed for treatment facilities. The hearing concluded on the note that a D classification was utterly impossible of achievement by industry, and that it was therefore unrealistic.

Following the hearing the Water Pollution Control Board took steps to solidify thinking about its proposed water classification plan. In so doing, it acted in accordance with certain basic principles. The first was that, in compliance with the law, it keep in mind always local usage. This admonition it interpreted in the widest possible way, taking account not only of local use of the Buffalo River, but of the whole background of the pollution reduction problem as well. This meant, among other things, that it gave due consideration to the abatement program designed and for years vigorously supported by the Buffalo Sewer Authority. A second principle was that it take into account the views of all interested parties. The result of observance of this rule was that representatives of the Board spent a great deal of time in conferences in and around the City of Buffalo during the several months following the public hearing. A third principle was that the Board refrain from precipitate action. It was in no hurry to adopt a final classification plan, but was content to let things simmer along until the time for adoption of a final plan seemed propitious.

It is well that the Board approached the local situation warily, for while there was wide support for the Sewer Authority's program, it was not viewed with universal approval. The most important doubter was the Erie County Health Department, which had raised questions, particularly regarding the cooling water project, some time before. The Department had taken advantage of its association with the Board on the latter's classification survey to remark, in November of 1952, that "this plan should be discussed further before accepting it as a partial solution of the Buffalo River problem." The introduction of Lake Erie water into the Buffalo River undoubtedly would have some beneficial effects, the Department continued, but it would not take the place of necessary pollution reduction work. Some months later the Erie County Board of Supervisors requested a statement on the subject of the County Health Department, and on April 13, 1953, the Department responded with a detailed statement on "Water Pollution in Erie County." Among other issues the report took up the cooling water plan, listing some arguments against it. Wholly apart from the cooling water project, the report concluded, industry must reduce its waste discharge into the River at least 50 per cent, and in addition, ". . . entirely remove materials that are particularly toxic to fish, such as oils, cyanides, sulphur compounds, etc." [17]

There was some feeling that, in view of the general support which the Sewer Authority plan commanded, the County Health Department was "off the reservation." The Water Pollution Control Board, seeking to get at the bottom of the issue, discovered that the Health Department was not unfriendly to the cooling water project, which, however, it insisted should not be confused with an industrial pollution reduction program. In a letter dated May 5, 1954, Frederick W. Crane, General Manager of the Buffalo Sewer Authority, sought to learn the attitude of the State Health Department regarding the Authority's pol-

---

[17] Erie County Health Department, *Water Pollution in Erie County*, p. 6.

lution abatement-cooling water project. The inquiry precipitated a conference attended by A. F. Dappert, Executive Secretary of the Water Pollution Control Board; Earl Devendorf, Director of the Bureau of Environmental Sanitation of the State Department of Health; and Dr. Berwyn Mattison, Erie County Commissioner of Health. These three agreed on a joint statement, which Devendorf enclosed in a letter to Crane dated May 28th. The statement applauded the cooling water project in three substantial paragraphs, then, in a one-sentence conclusion, voiced complete approval of the stand taken by the County Health Department:

> In view of the above, we view the proposed project as desirable, from the standpoint of over-all abatement of pollution, but only as a component part of the comprehensive program of pollution abatement which is envisioned for the future.

The statement was of special significance because of industry's emphasis on the pollution control aspects of the cooling water project. The industries were reluctant to accept the principle laid down in the joint statement, and indeed continued to press the argument that, once the cooling water project had been accomplished, their responsibilities for pollution reduction would have been discharged. The joint statement of May 28 indicated that that view would have rough sledding.

The above conference was but one of many participated in by representatives of the Water Pollution Control Board, principally A. F. Dappert, and, as a co-opted partner, Earl Devendorf, during the period November 1, 1953—October 1, 1954. It is important to remember that the Board was involved in classifying the waters of the entire Niagara Frontier. The Buffalo River represented but one aspect of the total problem, though because of its long history of gross pollution it lay near the center of the Board's concern. With respect to the Buffalo, it will be recalled, the problem of classification hinged on the question whether the River's water from the industrial plants to its mouth should be rated as Class D or Class E.

By the fall of 1954, Mr. Dappert was ready to make his recommendation to the Water Pollution Control Board. This he did in a letter dated October 18, 1954. The letter stated that

> the affected industries presented and filed very convincing evidence that it would be impossible and impractical to install sufficient treatment facilities to maintain a minimum of 3 parts per million of dissolved oxygen in the Buffalo River at all times . . . Industry presented, in my opinion, a firm and justifiable argument that the lower section of the

Buffalo River be assigned to Class E instead of Class D. This viewpoint was opposed by only one spokesman in behalf of conservation
interests at the October 21, 1953 hearing whose statements were of
a general highly critical nature but without support of any basic facts.
In view of the above, I am recommending Class E for the lower portion of Buffalo River even though I know . . . there will be some
criticisms raised if the Board so classifies Buffalo River.

The Board adopted this recommendation on December 6, decreeing
that it should become effective on December 15.

On January 13, 1955, the five industries received notice that the
water of the Buffalo River basin had been classified E, and immediately a round of conferences started on the steps necessary to bring the
water up to that classification. These conferences were both individual
(a representative of the Water Pollution Control Board conferring with
a particular industry) and general. A general conference was held on
February 25, at which time the industries presented an extended document summarizing their common view. They were satisfied with the
E classification, but appeared somewhat concerned about the abatement plan even then under development by the Board. More than that,
they were concerned about the relationship between the water classi-
fication/pollution abatement plan of the Board and the pollution abate-
ment/cooling water project of the Buffalo Sewer Authority. There
seemed to be general if tacit agreement that the Board's abatement
plan would supersede that of the Authority. There remained, however,
the troublesome residue of the cooling water plan, which seemed likely
to persist apart from any action the State might take. Let us leave that
issue for the moment in favor of further attention to pollution abatement as such.

The procedure adopted by the Water Pollution Control Board calls
for the drafting of a comprehensive pollution abatement plan (Step 5,
above) following formal adoption of the water classifications. Most of
the year 1955 was spent in the development of the comprehensive plan.
In keeping with standard Board procedure, the abatement plan was
developed in consultation with the industries and agencies to be
affected—indeed, the conferences called to effectuate the water classification and those held to discuss the developing abatement plan were
frequently the same, for measures taken to improve the quality of the
water naturally would be incorporated in the abatement plan in the
end.

By the end of the summer the Board's staff (principally again the
Executive Secretary, A. F. Dappert) was ready with recommendations

for action; and at the end of October (1955) the Board adopted a comprehensive plan for pollution abatement in the waters of the Niagara Frontier. Pages 64 to 70 of the plan contain brief descriptions of and detailed pollution abatement steps to be taken by Donner Hanna Coke Corporation, General Chemical Division of Allied Chemical and Dye, National Aniline Division of Allied Chemical and Dye, Republic Steel Corporation, and Socony Mobil Oil Company. In a letter of November 15, 1955, to Mayor Steven Pankow of the City of Buffalo, the five industries noted receipt of the comprehensive plan and stated that "The industries involved have now endorsed and are undertaking to comply with this Plan." [18]

As a conclusion to the comprehensive plan, the Water Pollution Control Board appended a discussion of the problem of enforcement. Emphasizing the fact that its own staff was quite limited, it requested full cooperation of the local authorities. The greatest part of the increased work load would fall on the Erie County Health Department, for which the Board bespoke additional personnel. The following recommendation is of special interest:

> With reference to water pollution from industrial sources, it would be very helpful to have created within the County Health Department a section for this specific work. This section should be staffed with Engineers who are familiar with industrial processes, sources of wastes in the various industries, and of sufficiently high calibre so as to command the respect of the plant managers. The section would work closely with the county laboratory in assisting these industries in locating and correcting sources of industrial pollution.[19]

This was not to deny the obligation of the Board to cooperate with local agencies in every particular, or to engage in enforcement work direct within the limits of its means. On the contrary, the Board promised that its representatives would proceed without delay to hold meetings with the industries and municipalities (in this case, the City of Buffalo) involved, with an eye to the inauguration of abatement

[18] The status of the Buffalo River problem as of the end of 1955 is summarized in the five-year progress report of the Joint Legislative Committee on Natural Resources, *Trends and Developments in Resources Conservation* (Legislative Document, 1956, No. 63), pp. 226-231. In particular, a useful summary of the conference of February 25, and of the stand taken there by industry, may be found on these pages.

[19] Water Pollution Control Board, *Comprehensive Plan for Abatement of Pollution of Waters of the Lake Erie (East End)-Niagara River Drainage Basins* (Albany, 1955). The discussion of the problem of enforcement appears at pp. 71-73.

programs through voluntary action. So well did these meetings go that, at the end of the year 1955, the Board was able to make the following report:

> Initial conferences have been held with many of these entities, with gratifying results, the purposes of which have been to reach definite agreements in each case as to the future steps to be taken to solve each pollutional problem. So far the manifested cooperation has been 100 per cent. The Board in all instances thus far has received commitments for what it has asked.[20]

And what of the water cooling project? The engineering survey sponsored by the five industries had resulted in a report in 1954 which in general confirmed the Symons recommendations of 1946 and in particular certified the Lake Erie water plan as to its engineering feasibility and financial soundness.[21] Industry had accepted and endorsed this report. In their statement prepared for the conference of February 25, 1955 (cited above), the industries summarized their view of the project. As they saw it at that time, there were three major problems. The first concerned the relationship between the cooling water project and a pollution abatement program. The industries continued to regard the water project as primarily a pollution abatement scheme, and to entertain the hope that, if that project should materialize, they might be held (largely if not wholly) to have discharged their pollution reduction responsibilities. As a matter of fact, they were more than a little annoyed that the scheme had come to be regarded as an industrial cooling water project, which they thought manifestly unfair. Second, the industries were interested to know what public agency would build and operate the project. They still favored the Buffalo Sewer Authority but were prepared to cooperate with the City of Buffalo if in the end it should be determined that the City should have the responsibility. Third, they sought an answer to the question, "Who is to assume the financial responsibility for the project?" The industries expressed themselves as being entirely willing to pay for the water they used, but as being both unwilling and unable to commit themselves to a long-term contract—the period most often mentioned was 25 to 30 years.[22] The attitude of the five industries was

---

[20] Joint Legislative Committee on Natural Resources, *Trends and Developments in Resources Conservation*, p. 247.

[21] This report grew out of the contract with Nussbaumer, Clarke, and Velzy undertaken as a result of the conference of September 19, 1952.

[22] Joint Legislative Committee on Natural Resources, *Trends and Developments in Resources Conservation*, pp. 229-231.

well summarized in the above-mentioned letter of November 15, 1955, to Mayor Pankow. Two excerpts will serve to point up their view:

> Each of the undersigned companies is therefore willing, so long as the operations of each of them shall necessitate the use of Lake Erie water for process cooling, to purchase said water at cost, . . .
>
> The industries would like to emphasize that this is primarily a pollution abatement project, and that their principal interest in it is its important bearing upon the achievement of the objectives of the Water Control Board's Comprehensive Plan for Pollution Abatement in the area.

On the other side of the question, the record shows no disposition on the part of any agency or official to doubt that the introduction of large quantities of Lake Erie water would have a beneficial effect on the Buffalo River. The joint statement issued in behalf of the State Department of Health, the Water Pollution Control Board, and the Erie County Health Department on May 28, 1954, conceded this without question. The beneficial effects, however, would be limited to (1) dilution of the polluted water and (2) increase in stream flow during the dry summer months. They would not extend in any important way to toxic materials, which must be withheld from the River at their source. This called for a pollution abatement program entirely separate from the cooling water program. The Water Pollution Control Board had indeed endorsed the cooling water project, as industry stated, but not as being in any sense the equivalent of or a substitute for a basic pollution control program.

By their ready acceptance of the Board's comprehensive pollution abatement plan, the five industries made possible the consideration of the cooling water project on its merits. The truth was, the industries wanted and needed the Lake Erie water. This was indicated by an official report of 1956, which contained the statement that "There is continued agitation by the industries for the Buffalo River Industrial Water Supply Pollution Abatement Project." [23] Industry was prepared, then, to proceed energetically with the negotiations which would be necessary to bring the proposal to a conclusion.

These negotiations, which were almost continuous throughout 1955, led to general agreement on the propositions that the project was a worthy one and that the City of Buffalo was the unit best equipped to undertake it. On December 13, 1955, Mayor Pankow recommended

[23] Advisory Board to the International Joint Commission on Control of Pollution of Boundary Waters, *Tenth Progress Report to the International Joint Commission* (October, 1956), p. 10.

to the Common Council that that body direct the Commissioner of Public Works to enter into negotiations with the industries looking toward the development of a contract. It was generally agreed that the City would require legislative authorization to undertake the project. The Joint Legislative Committee on Natural Resources assumed responsibility for that part of the problem.

The interminable conferences and discussions dragged on throughout the whole of 1956, with draft bills, agreements, and contracts shuttling among the interested parties in a steady stream. Progress, if slow, was inexorable, for the industries and the City had at last achieved a common ground of understanding.

The first positive break came when, in its session of 1957, the Legislature passed "The Buffalo River Improvement Act" (Chap. 493, Laws of 1957). The act's caption gives the gist of the measure; it reads: "AN ACT empowering the City of Buffalo to construct a waterworks system to supply raw water from Lake Erie for the purpose of relieving Buffalo River from pollution by sewage and wastes, authorizing it to lease said system, and authorizing it to levy taxes in order to pay for said system." In signing the act, on April 15, Governor Harriman commented that the agreement which it represented had been eleven years in the making.

The passing of the Buffalo River Improvement Act spurred contract negotiations, and by late spring of 1957 two documents had been drafted by the lawyers representing the City and the five industries. The first was an agreement, the second a lease; bearing the dates June 12 and June 13, 1957, they were judged to represent the legal action necessary to effectuate the cooling water project. In briefest summary, the documents represented mutual obligations undertaken between the City of Buffalo on the one hand and the Buffalo River Improvement Corporation, an organization representing the five industries, on the other. The City agreed to construct the cooling water project (at a figure estimated not to exceed $7,000,000) and to lease the facilities to the Corporation. On its part the Corporation agreed to make payments to the City sufficient to amortize the bonds issued by the latter for the project. The Corporation in turn would sell water to the five industries at cost, calculated on a basis which would yield an income sufficient to meet the payments due the City. The industries obtained a very important compromise, embodying a principle on which they had long insisted, in that any particular industry would be obligated by the contract for no longer than five years, with the agreement subject to renewal for subsequent periods of five years each up to fifty years.

Presumably the five industries would find these terms satisfactory. About the City there was some doubt; for while the Mayor was favorably inclined, the Council adopted a more cautious attitude. It soon became clear that there would be no early action. Then came one of those turns of political fortune which always occurs sooner or later: an election produced a new city administration, which in the course of time expressed its dissatisfaction with the five-year limitation of obligation granted the industries by the proposed contract. In subsequent exchanges both parties have proved obdurate, the City insisting on commitments for the full thirty years of the prospective bond issue, the industries holding out for commitments limited to five years but subject to renewal. Visual progress toward settlement of the dispute has been imperceptible. Negotiations nevertheless continue in the hope of reaching a compromise agreement, which is both the way and the expected result of administration.

CHAPTER VIII

Water Power: The Allocation of St. Lawrence Power

THE hydroelectric power potential of its many streams has been identified elsewhere as one of New York's principal water resources.[1] Governor Charles E. Hughes, in his annual message to the Legislature of January, 1907, called attention to this resource and recommended a strong legislative policy regarding it. Almost exactly half a century later, Governor Averell Harriman made public a decision which concluded action on an important phase of the State's fluid and still-developing power policy. The decision taken by the Governor (announced February 24, 1957) gave final approval to a basic element in the State Power Authority's plan for the allocation of power from the St. Lawrence hydroelectric plant then under construction. This chapter concerns the allocation of power from the St. Lawrence on the New York side of the river; it will treat in some detail of that subject, which lies close to the heart of the State's water power program.

The St. Lawrence is one of New York's two great "boundary rivers" (the Niagara is the other) which for a great many years have figured in negotiations and plans for hydroelectric development. The River drains a basin of more than 300,000 square miles, of which the Great Lakes are the central feature. The remarkable thing about the St. Lawrence, from the point of view of hydroelectric power generation, is its steady, even flow, which is guaranteed by the five great storage reservoirs in which it finds its source. Other rivers of comparable size experience maximum/minimum flow ratios of great magnitude: the Mississippi's flow ratio is 25:1, the Columbia's, 35:1. For the St. Lawrence, however, the maximum flow of record is only a little more than

---

[1] Chapter II, section on Water Power.

twice the minimum flow. From the point of view of power generation, the conditions prevailing on the St. Lawrence are therefore almost ideal; for the steady flow ensures that an unusually high proportion of all power produced will be "firm" power—power that can be relied on every hour of the day and every day in the year. It is not a matter for wonder that New York has eyed this vast power potential covetously for more than half a century.

On the one hand, then, the St. Lawrence long has offered rich rewards, almost, it has seemed, for the taking. Further, the physical problems have not appeared great: the Corps of Engineers, and more than one private engineering firm in addition, have certified the feasibility of the hydroelectric development of the St. Lawrence. Further still, the need for additional power in the area, long recognized, has increased substantially in recent years. But on the other hand, the institutional problems involved in the development have proved almost insurmountable—at least they have required half a century for their patchwork resolution. A general understanding of these problems is necessary as background for the chain of events on which this discussion focuses.

INSTITUTIONAL BACKGROUND

The complexity of the problems of developing the St. Lawrence for hydroelectric power production rose from three primary considerations. First, the St. Lawrence is an international boundary stream, and this of course meant that any development was dependent upon an agreement between Canada and the United States, the nations involved. Second, the government of the United States has a special interest in the St. Lawrence, primarily because it is a navigable river and secondarily because of the Federal power policy adopted in reference to navigable streams. Here, then, lay two complicating factors not normally present in the development of a state water program.

But third, New York compounded the complexity of the problem by its vacillation in developing a power policy. The State's dilemma and its gradual resolution have been sketched above (in Chapter II). It will suffice here, therefore, to pick up the thread with the passage by the Legislature of the Power Authority Act in 1931. From that year onward the State had an agency whose mission it was to develop and market St. Lawrence power. It was to be the lot of the Authority, in good years and bad, through depression and war, to negotiate continuously and to litigate occasionally for a long twenty-three years in its efforts to gain official approval of state development and marketing of New York's portion of St. Lawrence power.

International concern for St. Lawrence power roots solidly in a treaty entered into by Canada and the United States in 1909. In briefest terms, the Boundary Waters Treaty required that, before a project that would affect in any material way the waters common to the two nations might be undertaken by either, the approval of the International Joint Commission must be secured (Articles III and IV). The Treaty provided in some detail for the organization and jurisdiction of the International Joint Commission, which was to consist of six commissioners, three representing Canada and three the United States. The Treaty of course recognized and contained provisions regarding both navigation and power. Both are of unquestioned importance, but their linkage was to add greatly to the complexity of the power problem and to the delay in reaching a solution in the years to come.

There was considerable desultory action looking to the development of St. Lawrence power in the years following adoption of the Boundary Waters Treaty. An example is provided by a report issued by the I.J.C. in 1921 which dealt with the River's power potential and proposed facilities for their development. The first important break, however, occurred in 1932, when representatives of Canada and the United States negotiated a treaty which called for the construction of a 27-foot ship channel and, as a companion project, a power plant in the International Rapids section (the Barnhart Island site). Hope ran high that action might at last be imminent, but it oozed away month by month as the United States Senate debated the proposed treaty. In the end (mid-March, 1934) the Senate voted favorably on the treaty, but by less than the two-thirds majority (the vote was 49 to 43) required for approval. During the next several years sporadic efforts were made to revise the proposed treaty in a way to make it acceptable; but nothing tangible came of these efforts, and by 1940 it was generally accepted that the treaty of 1932 was a dead issue.

The protracted negotiations bore promising fruit nevertheless when, on March 19, 1941, the two countries entered into an executive agreement looking to navigation and power improvements on the St. Lawrence. Such an agreement would require for its implementation concurrent (or parallel) legislation by the national legislative bodies of the two countries, and this in its turn became a major stumbling block. First the onset of the war claimed the prior attention of both countries, then with the return of peace Congress manifested its usual reluctance to agree to joint international action. When bills which would have authorized cooperative action along the lines of the agreement of 1941 failed to pass both House and Senate Committees in the 82nd Congress (1952), Canada announced its withdrawal from the agreement and the slate was wiped clean once again.

Meanwhile still another approach, one which in its initial stages at least would not require congressional agreement, had been readied; and on June 30, 1952, Canada and the United States submitted identical applications to the International Joint Commission for approval of the construction of power facilities in the International Rapids section of the St. Lawrence. After some months of deliberations, punctuated by a number of public hearings, the Commission on October 29 approved the joint construction by Canada and the United States of the power facilities covered by the application. In contemplation of approval, Canada had already designated the Hydroelectric Power Commission of Ontario as its agent to act in the construction and operation of the Canadian part of the power project. It remained for the United States to take counterpart action.

Meanwhile the New York Power Authority was waiting impatiently in the wing for its cue to come onstage. The Authority indeed had been active from its inception in attempting to promote agreement on the development of St. Lawrence power; but its history had been a study in complete and utter frustration. Its strategy required that two important steps be taken in sequence: first, the Authority must contrive to have itself designated as this country's agent in constructing and operating any facility that might be agreed upon; and second, it had to procure specific legal authorization to proceed when the preliminary obstacles had been cleared.

New York's demand that it be permitted to develop St. Lawrence power was voiced early in the Senate deliberations on the proposed treaty of 1932. The State Power Authority and the Federal Power Commission joined to submit a summary of findings to the Senate in 1934. Any hope for immediately favorable action disappeared, however, with the Senate's failure to approve ratification of the treaty. From 1934 to the mid-'forties the Power Authority played a waiting if impatient game. During the war years it needled American authorities with occasional inquiries as to the state of the negotiations, and it corresponded and negotiated with opposite-number Canadian agencies to keep the issue alive. With the end of the war, numerous resolutions and bills were introduced in Congress to effectuate the 1941 executive agreement. The Power Authority appeared in a number of these measures, which however were uniformly dispatched either in committee or on the floor of the House or the Senate.

In 1948 the Power Authority, recognizing that it was making little headway in Congress launched new action in the form of an application to the Federal Power Commission for a license to develop St. Lawrence power. Once more the Authority was to meet with frustration; for on December 19, 1950 (a good year-and-a-half after its receipt), the Commission dismissed the application on the ground that it would be inappropriate for that body to take action on a matter pending before Congress. The Power Authority petitioned the Commission for a rehearing on its application early in 1951; and when its petition was denied, it sought action in the United States Court of Appeals to set aside the Commission's dismissal order and send the case back to that body for further study. Before the petition came up for hearing, Congress cleared the atmosphere by refusing to pass legislation which would have authorized United States participation in the executive agreement of 1941. Thereupon the Federal Power Commission joined the Power Authority in asking the Court of Appeals to dismiss the Authority's case and return the application to the Commission for further consideration. The request was granted, and on September 22, 1952, the Power Authority filed an amendment to its original application with the Commission.

Events were now approaching a climax; for while the New York Power Authority was pressing its cause before the Federal Power Commission, United States and Canadian representatives were busy with their application before the International Joint Commission for approval of a plan for joint construction of a power plant. As above noted, the latter application was approved on October 29, 1952, and this was a signal for the re-doubling of efforts by the Power Authority. From late 1952 until the spring of 1953, the Federal Power Commission investigated the Authority's application. Then the presiding examiner brought his deliberations to an end, and on July 15, 1953, the Federal Power Commission issued the Power Authority the long-sought license. On November 3, the Authority formally indicated its acceptance of the provisions of the license, and the principal substantive issue was thus resolved.

The way was not yet cleared for positive action, however, for announcement of the license brought forth a storm of protest. A number of interested parties, including the Lake Ontario Land Development and Beach Protective Association, the Central Pennsylvania

Coal Producers' Association, and the Public Water and Power Corporation of Trenton (New Jersey), petitioned the Federal Court of Appeals to nullify the license on the ground that the Federal Power Commission had exceeded its power in granting it. In January of 1954 the Court of Appeals upheld the right of the Commission to issue the license, whereupon the complainants took their case to the Supreme Court of the United States. That court sustained the validity of the license in a decision announced June 7, 1954.

Thus ended twenty-three years of futility and frustration. The Power Authority of the State of New York was in business, duly designated as the American agency for the development of St. Lawrence power and duly licensed to construct and operate power facilities jointly with its Canadian counterpart. Note should be made of the fact that, during the same period, Congress had authorized creation of the St. Lawrence Seaway Development Corporation to collaborate with a Canadian agency, the St. Lawrence Seaway Authority, in developing the St. Lawrence to a depth of 27 feet for navigation. Thus navigation and power development were permitted to go along hand-in-hand, in a manner befitting both.[2]

THE PROBLEM OF POWER ALLOCATION

Following formal acceptance of the Federal Power Commission's license, the State made a number of moves in recognition of the Power Authority's new status. First, the Governor reorganized the Authority, placing at its head Robert Moses, one of the country's best-known builders of public works. This appointment suggested that the State was ready to move fast in the execution of its construction plan. It also ensured that the Power Authority would suffer few dull moments

---

[2] The background data developed in these historical paragraphs rest chiefly on the annual reports of the Power Authority of the State of New York. The Authority's *First Annual Report,* Legislative Document (1932) No. 111 (Albany, 1932), presents the sequence of events from as early as 1900 to the end of 1931 in considerable detail. Particularly useful are the appendices, which include nineteen letters and a substantial memorandum which assert and argue New York's claim to prior consideration in any arrangement made for the development of St. Lawrence Power. There are letters to and from Governor Roosevelt, President Hoover, Secretary of State Stimson, and Power Authority Chairman Walsh. Subsequent reports are equally useful for material on special aspects of the subject. The Fourth Annual Report (1934), for example, contains several research bulletins and special reports on various technical aspects of hydroelectric power production and distribution.

thenceforward, and that it would not languish for want of public attention. The Authority moved quickly to establish new offices in New York City, to employ a full-time staff, to reorganize internally, and to come to grips with the financial and engineering problems whose solution was prerequisite to the beginning of construction. Major contracts were underwritten by a revenue bond issue of 335 million dollars which was marketed in December of 1954. The contractors began to move in, and by the end of 1954 preliminary construction operations in the vicinity of Massena were well launched.

The job of construction undoubtedly would be done well: Moses and his engineers would see to that. But this was only half the problem confronting the Power Authority, and it might well turn out to be the lesser half in terms of complexity and of time and energy required. The total amount of power available to the Authority would be about 940,000 kilowatts, of which 735,000 kilowatts were considered to be firm, the remainder interruptible or secondary. The marketing of this power, the other half of the Authority's responsibility, unquestionably would prove difficult, for it would involve the thorny issue of the allocation of a fixed (or at any rate limited) amount of power among a number of competing claimants. To this issue the Authority, through its newly established Division of Power Utilization, now addressed itself.

The Power Authority Act was unequivocal on the subject of allocation.[3] Enacted in response to vigorous assertion of a public power sentiment by Governors Smith and Roosevelt, the Act incorporated a strong statement of the public power doctrine. In particular, with reference to the matter of allocating the power to be developed from the St. Lawrence, it instructed the Power Authority (Sec. 1005, Sub-Sec. 5) that

> . . . in the development of hydroelectric power therefrom such projects shall be considered primarily as for the benefit of the people of the state as a whole and particularly the domestic and rural consumers to whom the power can economically be made available, and accordingly that sale to and use by industry shall be a secondary purpose, to be utilized principally to secure a sufficiently high load factor and revenue returns to permit domestic and rural use at the lowest possible rates and in such manner as to encourage increased domestic and rural use of electricity.

[3] The Power Authority Act, passed originally as Laws of 1931. Chap. 772, was incorporated in the Public Authorities Law (Chap. 870, Laws of 1939), as Article 5, Title 1.

Sub-section 5 contained additional specific guarantees to domestic and rural users, but the passage quoted constitutes the heart of the "preference clause." The interpretation and application of this clause was to keep the people of New York agitated and the Power Authority in hot water until well into 1957.

The purpose of the preference clause was clear enough in principle, but the problem of translating the principle into practice proved anything but easy. First, some argued that a doctrine given legislative statement in 1931 was not necessarily applicable under economic conditions which prevailed in 1955-56. Second, even if the principle be accepted (and most contestants were willing to accept it—in principle), what is the best way to serve domestic and rural users? Municipalities owning their distribution systems and rural cooperatives asserted a prior claim to electricity under the law; but if their claim be allowed, do they have the right to more electricity than they can show present need for? What policy should guide the Authority in reserving power against possible future domestic and rural needs? Apart from this kind of question, it was argued vigorously that existing private utilities themselves serve domestic and rural users, and that they therefore have a sound claim to additional power under the preference clause. Here indeed was a difficult problem for the Authority! Third, there was the troublesome issue (recognized in the preference clause itself) involved in the need to maintain the high load factor which alone would justify the low rates visualized by the Act. Ideally, the lowest rates would result from the complete utilization of all power produced, interruptible and secondary as well as firm, all the time. What policy would most nearly achieve this ideal condition? Fourth, the Authority early determined to limit the sale of power to the area within 150 miles of Massena (the point of production), on the ground that transmission for a greater distance would make St. Lawrence power as expensive as or even more expensive than private power. The decision on this basic point greatly restricted the range of choice available regarding allocation. Fifth, $22,000,000 yearly would be required to service the bonds which the Authority had issued, and this immediate and inflexible need hung like a cloud over the discussions of power marketing. Clearly economic imperatives might well require modification of politically and socially desirable ends. In any event, the Power Authority had to take into account a complicated set of economic factors in making its power allocations.

Sixth and finally, there was the knotty issue of vested rights. Regardless the Act of 1931 and its preference clause, there were certain existing producers and consumers of power which the Authority might find it both economically inexpedient and politically unwise to dispossess. Most important among these was the Aluminum Company of America, to whose problem the Power Authority turned first.

THE ALLOCATION TO ALCOA

As far back as 1903, the Aluminum Company of America had located a small aluminum reduction plant at Massena. Subsequently it had acquired a subsidiary, the St. Lawrence River Power Company, which had been chartered by the New York Legislature in acts passed in 1896 and 1898 to divert water from the St. Lawrence for the production of hydroelectric power. Alcoa's Massena operation prospered through the years to such an extent that, by mid-year of 1955, it had 6,000 employees and was by all odds the largest employer in the region. Its annual payroll was $30,000,000. Considering the state of the North Country's economy, the importance of Alcoa to the section can hardly be over-emphasized. Of relevance also is the fact that the company was the largest taxpayer in St. Lawrence County. Further, its subsidiary power company provided Massena with its municipal water supply, and was favorably regarded in the area. Alcoa, then, was firmly ensconced. Apart from all this, the good will of the company would prove highly useful in the disruptive days of readjustment to come. It was not without reason therefore that the Authority turned first to the Alcoa claim.

Alcoa's position regarding St. Lawrence power was not difficult to comprehend. It rested on three primary considerations. First, the company had been producing 65,000 kilowatts of power through its subsidiary power company; this source of power would be eliminated summarily by the new construction, which would inundate the old power facilities. Second, it had been purchasing 134,000 kilowatts of power from the Niagara Mohawk Corporation. This was high-cost power, in whose purchase, however, the company had been subsidized by the Federal government. This arrangement would terminate in 1963, at which time Alcoa would have to begin bearing the full cost of such power as it might still require of Niagara Mohawk. It is not recorded that the company made open objection to paying the market price for power, although it did point out that the absorption of the increased power cost into the price of its product would affect its competitive position. Third, Alcoa had been purchasing an additional 46,900 kilowatts of power on a month-to-month basis from the Ontario Hydroelectric Commission, a source which the company averred could not be relied upon. Alcoa's power requirements therefore were in the neighborhood of 246,000 kilowatts drawn from three sources, of which one was about to be destroyed, one would soon be comparatively high in cost, and one was uncertain. In the circumstances Alcoa took the view that, unless it could purchase a substantial amount of power from the Power Authority on reasonable terms, it would be unable to continue its Massena operations on a full-time basis.

Here the Authority for the first time confronted squarely the preference clause of the Act of 1931. Clearly Alcoa could not qualify as a preference user under the Act, yet equally clearly it would be what Robert Moses later called a "first-class tragedy" to allow the company to shut down in Massena after 50 years. The dilemma did not detain the Authority for long; it concluded quickly to enter into a power purchase contract with Alcoa, and conducted negotiations to that end. On April 9, 1955, the Authority revealed the details of the proposed Alcoa contract. The principal provisions subsequently controverted were three in number. First, the proposed contract called for the sale of a large amount of electricity, 174,000 kilowatts of firm power plus 65,000 kilowatts of interruptible power, to Alcoa. Second, the term of the contract was to be 48 years. Third, the contract provided that, in the event that Alcoa should be forced to shut down involuntarily (by "act of God" in case of a labor strike), the amount of power the company would be required to pay for would be substantially reduced.[4]

On May 10, the Power Authority held a public hearing on the proposed Alcoa contract.[5] Presiding over the hearing was Robert Moses, Chairman of the Authority, who immediately directed attention to Power Contract S-1, the proposed Alcoa agreement. The Chairman recognized Mr. Randall J. LeBoeuf, counsel for Alcoa, who presented the views of that company. Mr. LeBoeuf first reviewed the history of Alcoa's Massena operations, then summarized on the problems the

---

[4] *New York Times,* April 9, 1955, p. 6.

[5] Power Authority of the State of New York, *Report of Testimony and Statements Presented at the Public Hearing on the Proposed Contracts with Aluminum Company of America, Public Service Commission of the State of Vermont, City of Plattsburgh, United States Air Force* (Board of Estimate Room, City Hall, New York, N.Y., May 10th, 1955). While, as the title indicated, the hearing concerned four proposed contracts, most of the testimony taken related to the Alcoa contract, which is the center of interest of the present discussion. The remaining three contracts were the subject of a separate hearing held October 17, 1955. Except where otherwise indicated, the events recorded and views expressed in the remainder of this section rest upon the testimony developed at the May 10 hearing.

company faced in the prospective shifts in type and sources of power. He made the point that, even under the most favorable of circumstances, the company would be forced to invest over $25,000,000 to adapt its existing facilities to the utilization of new St. Lawrence power. In the company's view the new investment, together with the $91,000,-000 already invested at Massena, abundantly justified the long term and favorable rates offered by the proposed contract. Anticipating an attack on the involuntary shutdown clause, LeBoeuf defended the principle involved, noting that the company in any event would be relieved of no more than 50 per cent of its power-purchase commitment. On the positive side, the witness emphasized that Alcoa proposed to waive its claims to water rights and to dam site value in return for the proposed favorable contract. In conclusion, he voiced the judgment that the contract as it stood would enable Alcoa to continue in full operation at Massena. Apart from the company's official view, the record carries abundant testimony of the wide support enjoyed by the proposed contract among civic and other interested groups in the North Country.

Following the Alcoa presentation, a number of witnesses voiced objections to the proposed contract. First among these was a representative of the cooperative electric systems of New York, who viewed Alcoa's stand as nothing short of a request for state assistance to help the company gain an advantage over its competitors. The witness objected not only to the terms of the proposed contract, but also to what he termed the Power Authority's piecemeal approach to the allocation of power. He requested a thirty-day adjournment of the hearing so that his group might have experts study the proposed contract. Other opponents of the contract included the Chairman of the Public Power Committee of the National Union of Utility Workers of America, a representative of the United Automobile Workers, and the Legislative Representative of the Liberal Party, who echoed these arguments and introduced the preference clause in addition. Included in the record of the hearing are numerous statements, letters, and telegrams in opposition to the contract. The ultimate in opposition was voiced by Robert W. McGregor, Vice-President of the New York State Association of Electrical Workers, whose argument took a turn which caused Chairman Moses to inquire, "What do you want us to do, fold up?" To which Mr. McGregor replied, "Honestly, yes." Then he added as an afterthought, "You understand, nothing personal, however." [6]

[6] *Ibid.*, p. 88.

On May 14 (four days after the hearing), Senator Herbert Lehman wrote a letter to Robert Moses in which he recorded vigorous opposition to the contract as it stood. He objected to the term (48 years), which he regarded as something like three times as long as was prudent; and he questioned the basic wisdom ". . . of committing one-fourth of the power output of the St. Lawrence before an over-all marketing plan had been clearly laid out." The Senator concluded by calling attention to the clearly stated intention of the Act of 1931 to favor domestic and rural consumers. On May 17, Moses replied to Senator Lehman. Defending the proposed contract, he emphasized that the Power Authority was ". . . in no position to hit this company (Alcoa) over the head with an axe." The view that the Authority should have in hand a complete marketing plan before entering into any contracts he labeled a "manifest and patent absurdity." An early contract with Alcoa was absolutely essential, he maintained, if that company was to complete the new construction the contract would require of it in time to receive St. Lawrence power.[7]

On May 27, the Power Authority approved the proposed contract with Alcoa and sent it along to Governor Averell Harriman for his action. The contract as submitted to the Governor contained all of the controversial provisions: it provided for the sale of 239,000 kilowatts of power to Alcoa, it ran for 48 years, and it retained the so-called anti-labor clause. Under the law, Governor Harriman had 60 days in which to indicate his decision.

The political implications of the Alcoa contract had been clear for some time. They were sharpened by a front-page article which appeared in the *New York Times* on June 3 under the headline, DEMO-CRATIC FIGHT LOOMS OVER SALE OF STATE POWER. The article sought to array Governor Harriman and Mayor Wagner against Senator Lehman and Franklin Roosevelt, Jr., unsuccessful Democratic candidate for Attorney General. The argument was that Lehman and Roosevelt had taken an unequivocal stand for public power, whereas the Governor might, by approving the Alcoa contract, find himself placed in the position of "trimming" on the issue. The political pot boiled higher when representatives of thirty municipally operated power systems and rural cooperatives met in New York City and passed a resolution opposing the contract.[8] A few days later the State

---

[7] Both letters appear in *Ibid.*, Senator Lehman's letter at pp. 221-223, Mr. Moses' reply at pp. 338-341.

[8] *New York Times*, June 4, 1955, p. 18.

CIO appealed to Governor Harriman to reject the contract, terming it ". . . as obnoxious as the Dixon-Yates deal." [9]

Meanwhile, Governor Harriman had called a public hearing on the Alcoa contract. The hearing was held in Albany on June 14, with the Governor's Counsel presiding. Once again Randall LeBoeuf testified in behalf of Alcoa, presenting a vigorous defense of the contract. Among the several individuals who appeared in opposition were James C. Bonbright, a member of the Power Authority board in 1931 and subsequently its Chairman, and Leland Olds, assistant to the first Chairman of the Authority. The arguments for and against the contract were those which had been stressed at the May 10 hearing; except for a few changes in names, the June 14 hearing might indeed have been a recessed session of the earlier one.

The Governor now had the full facts before him. Rumors were rife concerning the action he would take. A week after the hearing, he spoke at Massena in a ceremony dedicating a new bridge across the St. Lawrence. There he adverted to the marketing of power, but carefully skirted the question of the action he would take on the Alcoa contract.[10] Meanwhile, the Governor's office continued to receive communications urging approval or rejection of the contract.

On July 1 the *New York Times* reported as definite a development which had been previously rumored: Governor Harriman had informed the Power Authority unofficially that he would not approve the Alcoa contract in its then form. He proposed a number of modifications, which the Authority undertook to negotiate with Alcoa; and on July 3 the Governor announced his approval of the contract as modified.[11] The modifications reduced the term of the contract from 48 to 43 years, and secured to the State the right after 25 years to re-examine the agreement as to half of the power then being supplied to Alcoa. Further, the controverted "anti-labor" clause was deleted from the contract. In his statement of approval, Governor Harriman once more called attention to Alcoa's economic importance to the region, and to what he regarded as the equities involved in the case. Further, he promised to ". . . continue to give full consideration in the distribution of St. Lawrence power to the needs of domestic and rural consumers . . ." [12]

[9] *New York Times*, June 14, 1955, p. 16.

[10] *New York Times*, June 23, 1955, p. 31.

[11] *New York Times*, July 4, 1955, pp. 1, 4.

[12] *Ibid.*, p. 4. This citation is to Governor Harriman's complete statement approving the Alcoa contract.

Not all were happy over the Governor's action. Some labor representatives and some spokesmen for public power kept up their attack, charging that the modifications effected in the contract were insignificant and that the agreement was not in harmony with the preference clause. In reply to these criticisms, Governor Harriman stated that

> The aluminum company has operated at Massena for more than fifty years and anyone who would use the development of St. Lawrence power to drive it from this area would render a profound disservice to the entire state and to the cause of public power in our country.[13]

Senator Lehman, while not entirely happy, issued a statement of approval in which he noted that the revised contract represented ". . . a vast improvement over the former proposed contract." To lay the matter properly to rest the *New York Times* added its benediction, opining ". . . that the Authority and the Governor have put first things first in assuring, as most important, the stimulation of the industrial economy of the area affected."[14]

POWER FOR REYNOLDS

If the decision for Alcoa was reasonably clear-cut, the next one proved considerably more difficult. The generous allocation to Alcoa left the Authority with more than 560,000 kilowatts of firm power subject to sale. The law required that preference be given to domestic and rural users, but there were not enough such users in the pre-determined marketing area to absorb such a quantity of power, nor would there be in the foreseeable future. The Authority could take care of the needs of all domestic and rural users in the North Country, set aside enough power to cover projected growth for ten years into the future, and still have almost 200,000 kilowatts of firm power for sale. Plainly what was required, the Authority concluded, was a new high load-factor industry for the region to absorb the surplus available and thus help keep domestic and rural power costs low, in harmony with the Act of 1931. The Authority set about the task of finding such an industry.

Several considerations complicated its search. First, the industry had to be a heavy and steady consumer of electric power. Second, it had to be one in whose operations cheap (seaway) transportation was an important factor. Third, it should be one to which proximity to immediate market was not of first importance. Fourth, it would be well if

---

[13] *New York Times*, July 5, 1955, p. 41.
[14] *Ibid.*, p. 28.

the industry were a big employer of manpower, otherwise a considerable part of the value of having it locate in the region would be lost. Two types of industry, the electro-metallurgical and the electro-chemical, qualified by most of these criteria. The Power Authority set in motion machinery designed to discover and attract to the North Country an industry to fit its needs.[15]

On May 26, 1955, the Trustees of the Power Authority passed a resolution instructing its officers to conduct negotiations with every company that manifested any interest in the purchase of St. Lawrence power. The State Department of Commerce joined in the hunt, as did a number of chambers of commerce and other business and civic associations. Distressingly few applicants appeared, and most of these did not seem overly anxious. The Power Authority revealed along in the spring that its staff was negotiating with two industries, the Reynolds Metals Company and the Allied Chemical and Dye Corporation.[16] The latter soon withdrew, however, leaving the field to Reynolds alone. That company was actively interested, but it suffered from a defect which threatened to prove fatal: the reduction plant which it proposed to locate at Massena would be a comparatively light employer, which would make it undesirable in terms of one important aspect of the region's economic need. It is in the fabrication of aluminum that high employment occurs, hence attention turned to the possibility of finding a fabricator to go along with Reynolds. That company did not wish to do any fabrication of its own at Massena, hence the proposition offered by the Authority did not appear particularly attractive. It was reported that Reynolds was considering another location.

For its part, the Authority continued to search for a companion arrangement that would use a large quantity of firm power *and* employ a large number of workers. Its efforts met with little success, and by the end of 1955 it was apparent that the wares the Authority offered had only a very limited appeal for industry. Meanwhile, time was growing short: precious months were slipping away—and there were those bondholders. Inquiry indicated that Reynolds, still desirous of increasing its reduction plant capacity, would be interested in resuming negotiations for the Massena location. There was to be one

[15] A publication by the Authority titled *Power Marketing* presents this phase of the story in attractive, highly readable fashion. The publication takes the form of a letter from the Chairman of the Power Authority to Governor Harriman under the date January 30, 1957. The letter was accompanied by a summary statement of the Authority's marketing plans and procedures.

[16] *New York Times*, May 28, 1955, p. 8.

difference: before, the Power Authority had insisted that Reynolds find a fabricator to complement its reduction plant; this time there was no stipulation regarding a fabricator, although both parties continued to recognize the importance of a fabricating plant in the total development. Reynolds more than once expressed its confidence that such a plant would be forthcoming, but was never able to translate its optimism into a commitment.

The year 1956 was not an easy one for the New York State Power Authority. The conversations with Reynolds went along well, but the issue of an accompanying fabricator remained unresolved. Other negotiations regarding allocation had to be carried along simultaneously, and these did not always proceed smoothly. Massachusetts and New Hampshire, long included in St. Lawrence power calculations in a vague sort of way, carried their cases to the Federal Power Commission when they found themselves outside the marketing area defined by the Authority. Both states withdrew their applications following an engineering study which upheld the conclusions reached by the Authority. The St. Regis Indians filed a suit against the State for $33,800,-000 on the ground that ownership of Barnhart Island rested with the tribe. The claims court before which the case came reserved judgment. The State Public Service Commission asserted its right to control the retail rates charged for St. Lawrence power in opposition to a like claim made by the Authority, causing Chairman Moses to accuse the Commission of making "wild, demagogic, and unsupported assertions." [17] Developments on the Niagara River power front required an ever-watchful eye on the part of the Authority.

Still it was a good year. Reynolds clearly meant business, and negotiations proceeded apace. The company talked of a plant to cost $88,000,000. Further, word got around that two automobile manufacturers, one of them General Motors, had expressed interest in the possibility of locating a fabricating plant at Massena. True, nothing definite had come of this talk, which nevertheless served to buoy up hope of an eventual satisfactory resolution of the dilemma. The terms of Contract No. S-5 had been agreed to by both parties by the end of the summer; and if there was nothing firm on the fabricator problem, the contract with Reynolds nonetheless represented a bird in hand. It called for the sale to Reynolds of 200,000 kilowatts of firm power and 39,000 kilowatts of interruptible power. The term of the agreement was thirty-three years. The Power Authority called a public hearing

[17] *New York Times*, February 29, 1956, p. 26.

on the proposed contract, to be held in New York City on October 10. As that date approached there was still no final word on the fabricator issue; hence the hearing was postponed until November 15, in the hope that something might be worked out in the meantime. Still nothing definite occurred, and the hearing went on as scheduled on November 15 without a commitment in hand for a fabricating plant.

Chairman Moses opened the hearing by causing to be read into the record two statements which anticipated much of the argument to follow.[18] The first concerned the oft-expressed fear of some municipalities that the Authority had not made adequate allowance for their future growth. He sought to give the municipalities full assurance on this point. The second came directly to the fabricating plant issue. The Chairman mentioned the extended negotiations which had been conducted on that subject, and expressed the opinion that ". . . these negotiations will before long reach a satisfactory conclusion." He voiced regret that there could be no guarantee given at the moment.

With this introduction, he called on Walter L. Rice, a Vice-President and Director of the Reynolds Metals Company. Directing attention to a forty-page statement which Reynolds had submitted to the Authority, Rice undertook to summarize his company's case. His principal points were these:

1. The use of St. Lawrence power for industrial development will promote the growth of northern New York, with great benefit to the region and its people.

2. The aluminum industry is particularly well adapted to the use of St. Lawrence power under the conditions under which it is available. Further, the growth potential of that industry is great.

3. "Reynolds is a dynamic factor in the aluminum industry."

4. Reynolds' proposed new plant at Massena will result in large expenditures both during the construction period and in operation, and in addition will provide permanent work for about 1,000 employees.

5. The terms of the proposed contract are "fair and equitable to both parties," though somewhat less generous to Reynolds than was the earlier contract to Alcoa. Mr. Rice developed each of these points at some length. With respect to the basic issue of a fabricating plant, he

---

[18] Power Authority of the State of New York, *Report of Testimony and Statements Presented at the Public Hearing on the Proposed Contracts with Reynolds Metals Co.* and others (Metropolitan Museum of Art Auditorium, New York, N.Y., November 15, 1956). The hearing concerned a number of other contracts as well, but attention here is centered on the Reynolds contract. Except where otherwise indicated, the summary which follows rests on this document.

was confident but circumspect. Recalling that Reynolds had originated as a fabricator, he recounted recent efforts to attract a fabricating plant to Massena and voiced optimism that a fabricator would be found soon. The tone of his statement was such as to recognize this as a problem central to the issue of contract approval.

The contract was supported by more than forty resolutions, memoranda, statements, and telegrams from a wide variety of units and organizations, most of them located in the North Country but some from outside the region. Among those represented were town and village boards, county boards of supervisors, chambers of commerce, service clubs, American Legion posts, the Grange, and a number of labor unions. The testimonials, though varied in source, were remarkably uniform in tone: they emphasized the need for the industrial development of the North Country, and voiced the conviction that approval of the Reynolds contract would signalize a long step in that direction. The location of Reynolds at Massena, they believed, inevitably would attract other industries. They spoke of the year-round employment which Reynolds would offer, and of the consequent stabilizing effect on the region's economy. They referred to the creation of new taxable properties, which were expected to be a great boon to the governments of the locality and so to its people.

That the contract should enjoy strong support from the region is hardly a cause for surprise, but that such support should be so nearly unanimous is perhaps occasion for note. Many influential voices from the vicinity were heard in support of the contract, almost none in opposition. As many as a dozen interviews conducted at random in and around Massena in the spring of 1957 failed to produce a discordant note: all persons interviewed were staunch Reynolds supporters. As many as three local government officials expressed regret that the Power Authority was forced by circumstances to favor industry over domestic and rural users; but, as one of them expressed it, "That's the way the ball bounces." The Authority has to meet its obligations to its bond-holders, and besides domestic users in the end probably will get all the electricity they need. And besides all that, all consulted agreed, you can't get around the fact that industrial development will be a great thing for the North Country. Some regarded Reynolds as an effective agent for keeping St. Lawrence power in the region, which they thought would be a good thing.

If the hearing produced vigorous support for the Reynolds contract, it disclosed sharp opposition as well. The first opponent to be heard was the President of the Rochester Gas and Electric Corporation, who expressed the view both that industry had been (or was about to be) allocated too much power, and that the small municipalities were receiving a larger proportional allocation than their needs would justify. He stated his position without equivocation in these words: "We want our share of power by any fair standard to be determined." [19]

Apart from this single instance, all vocal opposition to the contract came from representatives of municipalities, rural cooperatives, and consumer organizations (principally labor unions), all of whom may be classified as supporters of public power. The most telling argument was that advanced by the spokesman for the Municipal Electric Utilities Association, an organization of forty-two cities, towns, and villages engaged in the distribution of electricity. This individual, on pointed admonition by Chairman Moses (which brought sharp protests from other opposition speakers), telescoped his case into a few major points. These points, insofar as they related to the Reynolds contract, follow:

1. The amount of power allocated (and proposed to be allocated) to the aluminum industry, some 50 per cent of the firm power available, is too great to qualify under the "secondary purpose" clause of the Act of 1931 which alone justifies sale to industry.

2. The limitation of the market area of St. Lawrence power to the territory within 150 miles of Massena is arbitrary and indefensible. There is no sound engineering or economic basis for such a limitation, which automatically excludes a large part of the State from sharing in the benefits of the St. Lawrence development.

3. The favors being shown the aluminum industry, together with the narrowly delimited marketing area, combine to encourage monopoly rather than widespread use of St. Lawrence power.

4. The Power Authority Act has been re-enacted, with amendments, seven times since 1931. With each re-enactment the basic policy of the act, including the preference clause, has been reaffirmed. The Power Authority unquestionably has the legal power, as it has also the moral duty, to carry out the provisions of the act, including specifically the policy of giving preference to domestic and rural users.

The line of reasoning developed by the municipalities was buttressed by the attorney representing the five rural electric cooperatives of the State. Noting that his clients served about 5,100 rural consumers in twelve counties, he called attention to an application which had been filed in their behalf the preceding December 2. The rural cooperatives maintained that they had the right of first use under the law, and that

[19] *Ibid.*, p. 126.

the right of first use meant the right of first purchase. What they wanted, in brief, was the allocation for their use of 375,000 kilowatts, even though their current consumption was only 5,300 kilowatts. The nature of the argument may be seen from the following colloquy:

> Chairman Moses: Is it your contention that beyond what these people can use at the moment and pay for, that we are supposed to reserve for some future use a whole lot of power, irrespective of what we need to service our bonds? Is that your contention?
>
> Mr. Penberthy (attorney for the rural cooperatives): Yes, that is my contention, and it is my contention that you have the right, and I say the duty—
>
> Chairman Moses: How would we pay our bondholders?[20]

The Chairman pressed Penberthy hard on this point, and the latter defended his position vigorously, maintaining that the reputation of the Power Authority was in no wise linked to the aluminum industry, but that on the contrary "any banker in New York" would be glad to have bonds supported by the arrangement he suggested. For good measure, he argued that the Reynolds contract would cost the rural and domestic consumers of New York over $100,000,000, that being the amount by which he estimated consumer bills would be reduced over the life of the contract if the electricity proposed to be allocated to Reynolds were allocated to domestic and rural users instead, as he insisted the law required.

Apart from the public power organizations, a number of representatives testified against the contract in behalf of consumer groups. Chief among the latter were the International Ladies Garment Workers Union, the Textile Workers Union of America, the International Brotherhood of Paper Makers, the United Automobile Workers of the AFL-CIO, the Amalgamated Clothing Workers of America of the AFL-CIO, and the National Farmers Union. The burden of their testimony was to this general effect:

1. New York is a State in which the cost of electricity is high and consumption comparatively low. The St. Lawrence power development affords an opportunity to spread the benefits of low-cost electricity throughout the State. The policies of the Power Authority will not lead to this result.

2. Many industries have fled the State because of high power costs, leaving disadvantaged communities and under-employed populations in their wake. It is the obligation of the Power Authority to combat

[20] *Ibid.*, pp. 148-149.

this trend by making cheap power available over the broadest possible area.

3. The Authority has developed no over-all marketing plan, but from the pattern which is taking form it is clear that the end result will not be proper recognition of the preference clause of the Act of 1931.

4. The heavy industry which the authority has chosen to favor employs less labor in relation to investment than almost any other industry that might have been selected. There has been much talk of a fabricating plant to improve the employment outlook, but no commitment regarding such a plant has been secured.

5. The Authority has acted with undue haste in moving to allocate the power which has been entrusted to it. The problem is a complicated one and obviously requires more time for study and consideration than it has thus far received.

6. What is good for Reynolds Metals is not necessarily good for the consumers and the people of New York.

With testimony concluded and the hearing adjourned, the Power Authority Trustees retired with their 600 pages of record to weigh the pros and cons of the Reynolds contract. It was five weeks before they were ready to divulge their decision. On December 19, the Chairman announced that, by a three to two vote, the Authority had approved the contract. As expected, the agreement provided for the sale to Reynolds of 200,000 kilowatts of firm power plus 39,000 kilowatts of interruptible power. The contract term remained at thirty-three years.[21] The contract went without delay to the office of the Governor, where it was received without immediate comment.

The two dissenters were the Harriman appointees, former Governor Charles Poletti and Thorne Hills, a Lockport lawyer. Governor Poletti based his objection on what he regarded as the failure to give proper weight to the preference clause of the Power Authority Act, which as a member of the staff of the forerunner St. Lawrence Power Development Commission he had helped to draft. Hills voted against the contract because of the failure to turn up an accompanying fabricating plant. The votes by Messrs. Poletti and Hills symbolized the key issues in the controversy.

But if two of the Trustees were unhappy over the contract, the Chairman was not. Mr. Moses hailed the agreement in an ebullient statement emphasizing the importance of industrial development to the North Country and the prospect of the early acquisition of a fabrica-

---

[21] *New York Times*, December 20, 1956, p. 1.

tor.[22] His proclamation signalized the beginning of a spirited campaign designed to help the Governor reach a favorable decision. The *Watertown Daily Times* fired the opening gun in an editorial published the day after the Authority approved the contract. The editorial mapped out two courses of action: first, we have to get busy and find a fabricating plant, the editor said; second, every interested individual and organization must let the Governor know by every available channel of communication ". . . that the best interests of the St. Lawrence valley will be safeguarded if he approves the Reynolds contract." [23] The challenge was widely accepted, and groups all over the region passed resolutions supporting the Authority's action and urging the Governor to approve the contract. Newspapers editorialized and printed favorable "letters to the editor." Moses brought his vigorous pen to bear on the issue, and on January 30, 1957, addressed to Governor Harriman the letter-report on *Power Marketing* referred to earlier. Citizens to the number of almost 15,000 in and around Massena signed a petition requesting approval by the Governor, and the Massena Chamber of Commerce got ready to send a special train to Albany filled with townspeople to urge prompt and favorable action. (The special train project was abandoned on advice from members of the Governor's staff.)

The opposition forces likewise conducted a vigorous campaign. Representatives of a number of labor unions addressed a letter to the editors of the *New York Times* in which they went over the arguments, familiar by now, against the contract.[24] They commended the dissenting Trustees and called upon Governor Harriman to record disapproval. The Municipal Electric Utilities Association, in a meeting held at Syracuse, confirmed its opposition and sent a resolution to that effect to Governor Harriman.[25] The vote, however, was not unanimous, several municipalities feeling that, while the contract did not pay due respect to the preference clause, they could not afford to put themselves in the position of opposing the industrial development of the North Country.

Meanwhile, Governor Harriman had called for a public hearing to be held in Albany on February 6. The Governor designated his Counsel, Judge Daniel Gutman, as the presiding officer. Mr. Gutman took the precaution to request of Attorney General Louis Lefkowitz an opinion

---

[22] *Ibid.*
[23] *Watertown Daily Times*, December 20, 1956.
[24] *New York Times*, January 8, 1957, p. 24.
[25] *New York Times*, January 12, 1957, p. 20.

on the validity of the contract; the opinion rendered held that the contract did no violence to the preference clause. The Governor's hearing brought out almost sixty individuals, who required approximately nine hours for the presentation of their testimony. The record of oral testimony fills a volume of almost 350 pages, and there is a separate volume containing more than sixty resolutions, statements, letters, and telegrams. Both sides of the issue were presented vigorously and in detail, though the witnesses contributed little that was new to the argument.[26]

Walter Rice, of Reynolds, led off, as he had at the hearing of November 15. His main theme was the industrial development of the North Country and the part Reynolds was prepared to play in that growth. He was not in a position to make a definite commitment regarding a fabricating plant, and he made none. His stand (that is to say, the contract) was supported by the coterie of witnesses with whom we are by now familiar—spokesmen representing units of government, chambers of commerce, service clubs, and the like, most of them from the St. Lawrence region. An interesting variation was provided by the President of St. Lawrence University, who stated that the region's 4,000 college students, drawn from all parts of the State, were ". . . solidly behind the position presented here." He requested approval of the contract as an inspiring example of responsible official conduct that would be appreciated by the young citizens of the colleges.[27]

The opposition likewise produced familiar names and faces, with one or two significant exceptions. James C. Bonbright submitted a twenty-page letter in which he attacked the contract on both historical and economic grounds. Adolph A. Berle, Jr., representing the Liberal Party, appeared in person and presented a statement of the opposition argument. Mr. Berle identified himself as a member of the St. Lawrence Power Advisory Committee which had negotiated the 1941 agreement with Canada, an agreement which, though not ratified, nevertheless provided the basis for the arrangement finally hit upon. Berle attacked the contract on a number of counts, most of them familiar to one who has followed the controversy. As a positive suggestion, he favored the

[26] *Executive Hearing on Proposed Contracts for the Sale, Transmission and Distribution of Power by the Power Authority of the State of New York—with Reynolds Metals Co.* and others (Albany, Governor's Hearing Room, State Capitol, February 6th, 1957). The analysis of the hearing offered here rests on this record, except as otherwise indicated.

[27] *Ibid.*, pp. 108-109.

absorption of St. Lawrence power into the State's power pool and its dissemination thereby over the whole of the State. This, he maintained, would serve the interests of domestic and rural consumers much more effectively than the Power Authority's plan—which, he added, was a novel, not to say odd, invention of the last couple of years. This gave him an opportunity to level some counter-charges at Chairman Moses, which he did in language only a little less vigorous than that customarily employed by his adversary.

As the hearing drew to a close at the end of a long day, certain facts emerged. First, the people of the North Country, insofar as the record revealed, were almost solidly in favor of the contract. Second, there was no fabricating plant in sight. Third, the parties to the controversy were no nearer agreement than they had been when the hearing began. The situation was not well in hand, but on the contrary was quite sticky. Every procedural step short of action by the Governor had been taken; cleancut or no, the decision was now up to him.

What the Governor would do was a matter of wide conjecture. Pro-contract leaders in the North Country thought he would approve, for more than once he had spoken out in favor of the industrial development of the region. Further, as a Massena civic leader put it in conversation, "Harriman's heart has always been in the North Country. After all, his family came from Ogdensburg." Others were not so sure: the *New York Times* some days after the hearing reported persistent rumors to the effect that the Governor would not approve the contract.[28]

The dilemma confronting Governor Harriman was a real one. Time was short, and growing shorter. Reynolds was the only industry with the requisite qualifications that had ever manifested any real interest in locating at Massena; and it would be almost impossible to develop interest in a new industry, bring negotiations to a head, and complete construction by the time power would be available. So far as the prospect of a new industry was concerned, it was Reynolds or nothing. On the other hand, there was the matter of the missing fabricating plant. Further, Democratic governors for almost forty years had fought for cheap electricity for domestic and rural consumers, who many thought were given short consideration by the Power Authority's allocation plan.

Governor Harriman's action was prompt and decisive. First, he requested a Power Authority Trustee to sketch out an allocation which might be expected from the projected pooling of St. Lawrence and

---

[28] *New York Times*, February 10, 1957, IV, p. 8.

Niagara power.[29] From this sketch he concluded that a combination of the two sources into one would produce power in quantities which would service all reasonable needs of both industrial and domestic rural users in western and northern New York for many years to come. A relatively minor revision of the Reynolds contract would make it manageable on this score.

Second, he picked up his telephone and called the man who would know about the fabricating plant. A press release of February 13, 1957, tells this story:

> I have been in touch with Harlow Curtice, President of General Motors, with a view to encouraging General Motors to enter into a contract with Reynolds Metals Company for establishment of an aluminum fabricating plant at Massena. Mr. Curtice indicated his interest to me and the company agreed to undertake negotiations with Reynolds. I understand that negotiations between the two companies are going on now. I have no report as to whether an agreement has been reached.

The statement, if optimistic, was inconclusive. A second press release, issued later the same day, nailed the matter down. It read, in part:

> I am gratified by the proposal of General Motors Corporation to establish a foundry for casting of automotive parts at Massena, contingent upon the approval of the contract for the sale of power by the State Power Authority to Reynolds Metals Company. This contract for the sale of power is now before me.
>
> The Reynolds Metals Company has agreed to a modification in the proposed contract relating to the rates to be paid for power when and if a tie-in of St. Lawrence and Niagara power is effected. With this change it is my intention to approve the Reynolds contract.

Thus were removed the two major obstacles to final approval of the Reynolds contract. Daniel Gutman, Counsel to the Governor, confirmed the significance of the events of February 13 in a press release issued the following day. "The determining factors in (Governor Harriman's) decision," Mr. Gutman stated, "were the addition of the General Motors plant, which will bring added employment to the area, and which is expected to expand substantially over the years, and the change in the contract agreed to by Reynolds."

Subsequent events were anti-climatic. Shortly after the announce-

[29] The Power Authority Trustees on February 4, 1957, had adopted a resolution calling for the merging of the market areas of the two projects. The resolution outlined an "integrated power marketing policy," to which the Power Authority pledged itself.

ment of February 13, representatives of the interested parties met and worked out the details of the proposed changes in the contract. Some days later, Governor Harriman announced that he would approve the agreement as amended; and on February 24 he released a statement indicating his formal approval of the contract.[30] Thus ended the most important episode in the history of the allocation of St. Lawrence power.

DOMESTIC AND RURAL ALLOCATIONS

The Power Authority Act by clear implication divided the consumers of hydroelectric power into two categories, industrial and domestic/rural. The Power Authority accepted this classification and, according to its lights, made allocations in harmony with it. The contracts discussed thus far allocated to industry somewhat more than half of the firm power available. While these contracts were under negotiation, the Authority began work on arrangements for channeling the remainder of its firm power (a little less than half of the total) to domestic and rural consumers. The negotiation of the domestic/rural contracts proved nothing like as harrowing as the task of working out the industrial agreements, particularly the Reynolds contract; for generally speaking the proposed domestic allocations were clearly within the spirit of the act of 1931, hence their negotiation rarely involved fundamental issues.

There were two major exceptions to this generalization. First, the Power Authority Act provided that hydroelectric power projects should be operated primarily ". . . for the benefit of the people of the state as a whole. . . ." The limited market area adopted by the Authority, by restricting the sale of power to a relatively small and sparsely populated section, seemed to many to preclude the enjoyment of the benefits of cheap power by the "people of the state." Public power advocates, including chiefly the representatives of labor unions and spokesmen for the Liberal Party, hammered away at this theme whenever and wherever they found an opportunity. They wanted St. Lawrence power to be incorporated into the "state power pool," and so distributed broadly throughout the State. The Power Authority stood firm on the market area as originally defined.

Second, there was some difference of opinion on the question whether the Authority should utilize existing distribution systems, even though private, or build its own transmission network and sell power direct

[30] *New York Times*, February 25, 1957, p. 1.

to local distributors. So far as the record shows, the Power Authority never seriously considered building its own transmission system, although some advocates of public power insisted that it should. Instead, the Authority decided early to utilize existing distribution facilities, and with one exception (the transmission line from Massena to Plattsburgh) it pursued this policy throughout.

During the spring of 1955 contracts were drafted with three prospective domestic/rural consumers: the State of Vermont, the City of Plattsburgh, and the United States Air Force Base at Plattsburgh. The contracts proposed to allocate to the three 100,000, 30,000, and 10,000 kilowatts respectively. A public hearing on the three contracts was held on May 10, 1955, in conjunction with the hearing on the Alcoa contract (see footnote 5, above). Spokesmen for the three contractors had little difficulty making the case that their principals served domestic and rural consumers and so qualified under the law. Opposition came nevertheless from a representative of the State Association of Electrical Workers, who pointed out that the contracts did not call for the employment of union men nor for the payment of union wages in construction, maintenance, and operation work. He did not believe that the government, as represented by the Power Authority, should be in the power business at all.

A second and much more spirited controversy arose over the issue of who should build the transmission lines. There were no lines of a size adequate to carry the contracted power from the plant at Barnhart Island to Vermont, Plattsburgh, and the Air Force Base. The draft contracts called for construction by the Power Authority, and the hearing developed strong support for that plan. Representatives of the Niagara Mohawk Power Corporation and the New York State Electric and Gas Corporation voiced vigorous opposition, however, proposing instead that they furnish the transmission facilities. On request by a number of interested parties, who felt that so controversial an issue could not be resolved in so short a time and who maintained moreover that the problem required further study, the Chairman recessed the hearing subject to later call. The recessed hearing re-convened on October 17, when the controversy resumed.[31]

[31] Power Authority of the State of New York, *Public Hearing upon the Terms of Proposed Contracts for the Transmission, Distribution and Sale of Power to be Generated at the Barnhart Island Power Plant of the St. Lawrence River Power Project to the Public Service Commission of the State of Vermont, The City of Plattsburgh, N.Y., and the United States Air Force for Its Base near Plattsburgh, N.Y.* (Hearing Room "C", 270 Broadway, New York, N.Y., October 17, 1955).

The Authority meanwhile had gained the support of Governor Harriman in building its own transmission lines to the East. The hearing failed to produce an agreement, and shortly after its adjournment a spokesman for the New York State Electric and Gas Corporation stated that his company would carry the case to court, if necessary, to win the right to transmit St. Lawrence power.[32]

On November 10, the Trustees of the Power Authority by a vote of three to two approved a plan for construction of the transmission lines by the Authority. Not entirely satisfied with this decision, the Trustees at the same time authorized the Chairman to attempt to arrive at a compromise arrangement with the private utilities. This attempt proved successful, and on December 7 the Trustees, at a meeting held at Massena, approved a revised plan by which the Authority would assume responsibility for the main line and the utilities for the local connecting lines.

On December 12, the Authority sent the three revised contracts to Governor Harriman for his approval. On January 17, 1956, the Governor announced his approval, and one week later the Power Authority entered into a formal agreement with the Niagara Mohawk Power Corporation and the New York State Electric and Gas Corporation to give effect to the agreement reached concerning transmission facilities.[33] Thus did the Authority contract for sale to domestic/rural suppliers of 140,000 kilowatts of firm power.

Chief among the private utilities active in the North Country was Niagara Mohawk, which in 1955 served 537,000 of the 600,000 domestic and rural consumers in the St. Lawrence market area.[34] The Power Authority proposed to do business with Niagara Mohawk, and to that end negotiated a contract which received consideration along with the proposed Reynolds contract. It pledged the Authority to furnish the company 115,000 kilowatts of firm power. On its part, the company agreed to purchase any amount above that figure that the Authority might have for sale, thus, its spokesman averred, guaranteeing the success of the power enterprise. The company also agreed to pass on to its domestic and rural consumers any saving that might result from its purchase of St. Lawrence power,[35] and to transmit for the account

[32] *New York Times*, November 3, 1955, p. 28.
[33] *New York Times*, December 15, 1955, p. 28; January 18, 1956, p. 25.
[34] *Power Marketing*, p. 18.
[35] With the December 1, 1958, bill each customer received a letter from company President Earle J. Machold reciting that "With this electric bill, Niagara Mohawk is pleased to start passing on to you savings realized on the power we

of the Authority any St. Lawrence power that might be purchased by any municipality or other public distributing agency in the St. Lawrence market area.

The Niagara Mohawk contract came up for consideration at the public hearing of November 15, 1956 (see footnote 18, above). Attention centered on the Reynolds controversy, although the advocates of public power, including specifically representatives of the Municipal Electric Utilities Association and of the rural cooperatives, had opportunity to record their opposition to the Niagara Mohawk contract. The basic objections were two: first, on ideological grounds, that the Power Authority should market its power direct; second, that the Authority did not propose to allocate enough power to municipalities and cooperatives to take care of their future needs. The Authority over-rode these objections and recommended to the Governor approval of the Niagara Mohawk contract. At the Governor's hearing of February 6, the November 15 performance was repeated all over again. Niagara Mohawk appeared to emerge from the hearing in good shape. Some days later it developed that Governor Harriman wished to suggest a modification in the contract terms which would secure special consideration to domestic and rural consumers when the anticipated tie-in of St. Lawrence and Niagara power should become effective. On February 17, representatives of the Power Authority and Niagara Mohawk met with the Governor and his Counsel to discuss the proposed changes in the contract. On February 24, Governor Harriman, indicating his satisfaction with the revised contract, announced that he would give the document his formal approval.[36] Thus ended another important phase in the history of power allocation.

The New York State Electric and Gas Corporation served some 21,000 domestic and rural consumers in the St. Lawrence market area, principally in the northeastern part of the State. The Authority elected to do business with the Corporation, and so negotiated a contract on which a public hearing was held on February 4, 1957.[37] As had been their

---

have purchased from the new St. Lawrence River Power Development, a project of the Power Authority of the State of New York." The first savings, the letter continued, amounted to $240,325, which was distributed to domestic and rural consumers on the basis of 1.37 of a mill per kilowatt-hour used.

[36] *New York Times*, February 25, 1957, p. 1.

[37] Power Authority of the State of New York, *Report of Testimony and Statements Presented at the Public Hearing on the Proposed Contracts with New York State Electric and Gas Corporation* and others (Port Authority Building, New York, N.Y., February 4, 1957).

custom, the Municipal Electric Utilities Association and the five electric cooperatives opposed the contract, the former on the ground of an excessive allowance for wheeling, the latter because it objected to the Authority's method of distributing its power. The contract nevertheless was approved by the Authority, and presently by the Governor as well. It called for the allocation of 20,000 kilowatts of firm power to the Corporation.

The public distributors of power to domestic and rural consumers in the market area included sixteen municipalities (excluding Plattsburgh) and three rural electric cooperatives. Through their respective associations they had maintained the view all along that, as public bodies serving consumers direct, they should be given preferential treatment. Consequently they had objected to the Power Authority's marketing plan at almost every turn as Chairman Moses observed testily in his "Madison Avenue" report, and had given that agency a hard time.[38] In addition to general objections on a number of specific points, in which the two groups often joined, the rural cooperatives proposed a plan by which they would become custodians of half of all the firm power, even though their demand at the time was only a matter of some 5,300 kilowatts. Notwithstanding what it appeared to regard as the reprehensible conduct of these public bodies, the Authority reported that about 70,000 kilowatts had been allocated to them—enough, it averred, to meet their needs for at least ten years.[39]

The accompanying table presents a summary of the Authority's allocation of firm power. It indicates that considerably more than half of all such power was allocated to three major industries, a good deal less than half to distributors to domestic and rural consumers. The only St. Lawrence power sold outside the State will go to Vermont, which will receive somewhat less than 14 per cent of all firm power generated. A majority of the local consumers will be served by private utilities: Niagara Mohawk alone will receive almost 16 per cent of the firm power sold, local public bodies (municipalities and rural electric cooperatives) only 14 per cent.

Thus did the State Power Authority discharge its obligation to market the power accruing to New York from the St. Lawrence. Many defend with vigor the Authority's allocation program, others attack it as a

[38] *Power Marketing*, pp. 18-19. The report was so characterized by a spokesman for the municipal electric utilities group.

[39] *Ibid.* Actually, the projected tie-in with the Niagara development subsequently made it possible for the Authority to enter into 25-year contracts and to undertake an increase in the contract rate of delivery during that period.

TABLE 3

Allocation of Firm Power from the St. Lawrence
Contract Status, December 9, 1957*

	Kilowatts		Percentage Distribution	
Industrial: all contracts executed				
Reynolds	200,000		27.2	
Alcoa	174,000		23.7	
General Motors	12,000		1.6	
Sub-total, industrial		386,000		52.5
Domestic/rural				
Contracts executed:				
Niagara Mohawk Power Corporation	115,000		15.7	
State of Vermont	100,000		13.7	
City of Plattsburgh, N.Y.	30,000		4.1	
New York State Electric & Gas Corp.	20,000		2.7	
Village of Solvay, N.Y.	12,000		1.6	
Plattsburgh Air Force Base	10,000		1.4	
Village of Boonville, N.Y.	3,800		.5	
Village of Rouses Point, N.Y.	2,000		.3	
Contracts approved by Power Authority Trustees:				
Village of Philadelphia, N.Y.	800		.1	
Village of Theresa, N.Y.	600		.1	
Contracts under negotiation:†				
Various municipalities and rural electric cooperatives (3)	54,800		7.3	
Sub-total, domestic/rural		349,000		47.5
Total	735,000	735,000	100.0	100.0

* Adapted from a typescript table furnished by the State Power Authority, December 9, 1957.

† Niagara Mohawk has agreed to purchase this power until it is needed by the municipalities and rural cooperatives.

subversion of declared public policy; but few would care to argue that the plan does not incorporate sound business principles. Good business has been the preoccupation of the Authority, and especially of its Chairman, all along. There is no reason to suppose that the primary goal of financial soundness has not been achieved. Whether financial soundness should have been set as the primary goal is another question.

CHAPTER IX

New York's Water-Resource Agencies in Action

THE foregoing accounts are designed to show how New York's principal water-resource agencies operate in on-the-ground administrative situations. Developed without embellishment and without conscious attempt to guide thinking, each case is intended to reveal how a particular problem was resolved under conditions which prevail in administration, if not normally at least frequently. It is fair now to ask the question, what is the significance of the cases, in the aggregate, for the observer of the administrative process? What lessons do they afford one interested in learning how New York manages its water resources? In answer, they give rise to two kinds of generalizations. One category, drawn direct from the cases, concerns the nature of administration;[1] the second, either taken direct or distilled from the cases by inference, consists of an evaluation of the State's organization for water resources management.

THE NATURE OF ADMINISTRATION AS REVEALED BY THE CASES

From the four cases, reviewed systematically in terms appropriate to administrative analysis, a number of general hypotheses on the nature of administration emerge. One proposition, a basic one, concerns the

[1] It will be understood that "administration" in this setting normally means water-resources administration. The hypotheses to be noted grow from case situations involving water-resource agencies in water-management contexts. Even so, the question whether, administration being a stage in the process of government which inheres in all programs, the generalizations gleaned here are not of universal or at least of wide validity is worthy of speculation. The examples are limited to one action area, but their lessons may not be.

significance of the physical, social, economic, and political setting—in short, of the total environment—for administration. It is commonplace that administration takes place in a real world; its place of action is a stage peopled by flesh-and-blood characters, among which of course, but by no means alone, are the administrators themselves, To say this is to allow that many factors enter into an administrative decision in addition to, and indeed apart from, the "merits" of the case. Among these are such considerations as America's traditional distrust of government, which pretty well ensures that, except in time of emergency, every important official proposal will be looked at inside and out and no precipitate action will be taken. Some critics characterize this innate spirit of caution as inertia, and identify an "administrative lag" which makes certain that government action normally will follow social need by some years. Another environmental factor is found in the American tradition of individual rights, which requires that every person with an interest in a public proposal be given a chance to be heard before action is taken. One aspect of this doctrine assumes the form of our characteristic faith in free enterprise, which made the five major industries such potent forces in the Buffalo River case. Private enterprise in turn is translated into vested interests, prominently represented by the Aluminum Company of America and the Niagara Mohawk Power Corporation in the St. Lawrence case. The under-developed economy of the North Country was cited frequently both as reason for the development of St. Lawrence power and as justification for the allocation policy pursued. The location of the Buffalo River in a heavily populated metropolitan area made it imperative that pollution abatement measures be taken. The cases offer other examples of the importance of environment, which serves administrative action variously as stimulant, guide, and brake.

In another direction, the cases offer ample testimony of the complexity of administrative action. Many straightforward measures which deal with non-controversial matters naturally enough are taken without the hubbub described above. Almost any significant action, however, is likely to be made the subject of challenge. In this event the original issue gets multiplied by a factor of, shall we say, ten, and the Black River Regulating District, to illustrate, finds itself arguing questions only indirectly related to its central proposal. In the process, the agencies and units of government involved also multiply, with new organizations asserting their concerns as the issues broaden. So do the private parties and interests concerned. What begins as a seemingly simple proposal—to clean out the Buffalo River, for example—thus grows into a complicated procedure involving ten or more administrative agencies, all three levels of government, and five major industrial plants. The point is not that such complexity may not be foreseen; on the contrary, it is a quite normal accompaniment of an important administrative action. The point is rather that complexity is the essence of administration, as the four cases abundantly illustrate.

Much has been said about the inter-agency and intergovernmental character of administration. Each of the four cases provides ample evidence on this point, and one, the St. Lawrence case, adds a generous dash of international involvement as well. From the general proposition that administration is a complex business, the secondary observation that administrative action frequently cuts across agency and unit lines, and sometimes international boundaries as well, follows almost as a corollary.

Yet another semi-corollary of complexity is found in the conclusion that the core of administrative action is found in the engineering of consent. Fiat action on an important issue is rare in administration; at least the four cases examined here provide no illustration of such action regarding a primary decision. The engineering of consent in turn is of the essence of politics, which suggests that administration and politics are not the mortal enemies some have professed to believe them. As a matter of simple fact, the two are very closely related, and may be considered reverse sides of the same coin. The Chairman of the Power Authority, widely regarded as an appointive executive who operates above the din of political battle, found it necessary to win over a considerable number of opponents in arriving at an allocation of power from the St. Lawrence. In the simplest and most straightforward of the four cases, that involving the Cannonsville reservoir, the New York City Board of Water Supply deemed it advantageous to proceed with its property acquisition program through personal negotiation rather than through direct legal action. The reputation for fair trading and good citizenship which is expected to accrue from this policy will stand the Board in good stead in the years to come.

From the foregoing it would seem to follow that administration is in good part negotiation, and this is indeed the case. Such negotiation may be conducted largely behind the scenes, among the administrative agencies and parties directly involved in a proposed action, or it may be taken to the public forum with the avowed purpose of gaining supporters for a proposed course of action. A "newspaper war" is a frequent companion of an administrative action. From this it follows further that a significant component of administration is found in extra-official relations, with client/consumer groups, with interest (pressure) groups whose concern may or may not be substantial, with the public at large. Each of the cases abundantly illustrates the significance of such relations for administration. The Black River Regulating District Board, to illustrate, found itself involved in official negotiations with the highest officers of the State, including the Governor, and with federal agencies as well. Official relations apart, it consulted continuously with its immediate clients; fought a constantly expanding war with what it considered to be outside groups; prosecuted and defended its cause in court and eventually carried its case to the public in two spirited if ineffectual statewide political campaigns. Here is an extreme example of the relations, official and extra-official, which administrative agencies and administrators must be prepared to pursue.

These observations will lead the thoughtful reader to ponder the relation between the public interest and private interests. Did the Power Authority's recognition of Alcoa's prior claim to power further or affect adversely the public interest? The same question is relevant to its decision to make a large block of power available to "preference" customers through the facilities of the Niagara Mohawk Power Corporation. Further, what of the decision to go out and find a new high load-factor industry for the North Country? Here were three basic decisions, each on its face favoring an important private interest; were the decisions at the same time taken in the interest of the public? Nor is the clash limited to that between public and private interests, for frequently there are two or more competing publics involved in a proposed action. In the Cannonsville case, to illustrate, the decision of the Board of Water Supply to go to the Catskills for water involved the City of New York in a conflict of interests with (rural) Delaware County. In such a conflict, where do the equities lie? Is the larger and more powerful public interest necessarily the greater? These and many like questions concerning the nature of the public interest present themselves for consideration. It is clear that they cannot be answered here, though the four cases afford many interesting points of departure for speculation.

Much of what goes before suggests the difficulty of keeping negotiations (and the consequent discussions) focussed on the issue at hand. The Black River Regulating District Board started out to build a dam, but soon found itself charged with the intent to attack if not destroy the Forest Preserve. A fly-casting enthusiast, seizing upon the opportunity afforded by the proposal to clean up the Buffalo River, advocated that the River and its tributaries be restored to their earlier condition as wilderness trout streams, at whatever cost to industry. The New York City Board of Water Supply found its program assailed by a local citizen who invoked the symbol of the Vanishing Redskin in his attack on the slickers from the city. To conclude that these and countless other minuscule or foreign issues were not relevant to the central proposition is not to imply that they were not important to some people in the context of the proposed action. It is only to suggest that tangential questions frequently insinuate themselves for consideration, often at the expense of judicious consideration of the main problem and sometimes to the utter exclusion of reason. It is, in short, to suggest the presence and something of the nature of irrational factors which affect and sometimes determine administrative action.

Almost all of the discussion to this point attests the informal, the tentative, and in the end nearly always the compromise character of administrative action. This results in part from the nature of the legal authorization, which usually not only permits but requires interpretation in shaping policy and giving it effect; in part, from the consequent presence of tenable alternatives; and in part from the constant need to reconcile conflicting interests. Consider, for example, the relationship between the pollution abatement program for the Buffalo River and the cooling water project for the five major industries located on the River. Although the two were essentially separate, the industries were so successful in tying them together that they were able to exact public assistance in the cooling water project as the price of peaceful participation in the pollution abatement program. As another example, the allocation of St. Lawrence power is hardly what one would have anticipated from reading the law; clearly large elements of interpretation and compromise entered into this decision. What a statute provides and what administration brings forth in the end are not always or necessarily wholly congruent, for the flexibility which makes administration work also ensures that a particular program oftentimes will turn out in practice to be something more or less than, and maybe something different from, what the law seemed to promise.

Adverting to the complexity of administration, it is clear that any particular administrative action is likely to take place over a long time: the four cases examined above required anywhere from ten to twenty-five years for their resolution. Explanations for this time lapse may be found in the requirements of democratic procedure, concerning which New York law is quite explicit; in unabashed red tape, which may be translated as unduly involved procedure; in the complicated nature of the problems addressed; in inadequate governmental machinery; and in the absence of vigorous administrative leadership. The culprit may be found in one or more, or in a combination of all. Whatever the explanation, it appears that administrative action is more likely to be long drawn-out than precipitate. The case for movement normally has to be constructed carefully step by step, but the brakes are built-in.

It follows from the foregoing that an administrative decision, taken frequently over a considerable period of time, often encounters changing conditions while in process of formulation. Thus action appropriate to the set of circumstances prevailing at one time during the deliberations may be quite out of keeping with the conditions obtaining at another time. The interpretation given the preference clause in the allocation of St. Lawrence power is believed by many to have been contrary to the Power Authority Act passed twenty-five years before. Conditions obviously have changed since 1931; the question is whether they have changed enough to warrant the new interpretation of the preference clause adopted by the Power Authority in 1955-1956. We do not have to answer this question to agree that, in words of ancient wisdom, circumstances alter cases, in the field of administration as elsewhere.

Finally, the four cases provide bases for speculation regarding the role of special agencies and districts in administration. It is significant that no agency in any of the cases saw, because none was commissioned to look, beyond the limits of a particular job defined in narrow, specialty terms. The Board of Water Supply wanted a dependable source of good water for New York City; it regarded any other purposes that might have been served by its Cannonsville reservoir as both incidental and comparatively unimportant. The Board of the Black River Regulating District sought only a means of regulating the flow of the River; again, other possible purposes were considered inconsequential. The Power Authority strove single-mindedly to market the power to be produced at Barnhart Island, looking neither to the right nor to the left for secondary purposes in want of sponsorship. Each agency performed with zeal, if not with uniform success, the single major function for which is had been created. But what of the many programs which command no such energetic champions? And what of the collateral activities, mayhap equally worthy, which might have profited from association with these single-purpose agencies? What, indeed, of the special district/authority versus the general government approach to water resources administration? The four cases provide illustrations of a number of single-purpose agencies in operation; they offer no clue, excepting through default, to the possible values of a multiple-purpose approach to the same problem areas.

A Critique of Administrative Structure

It remains now to appraise New York's organization for water resources management in more general terms. It is fitting that such appraisal should be undertaken at the end of a section on administrative action; for it is generally true, in administration as elsewhere, that an organization is to be weighed by its performance. Not every point to be made here stems from the foregoing case studies of administrative action, but most do; and all flow, either directly or by inference, from what has gone before.

One of the outstanding characteristics of New York's system of administration arises from the fact that the executive departments (nineteen of them) are enumerated in the Constitution, and that the Legislature is forbidden to establish additional departments. This limitation does not constitute an absolute prohibition on administrative reorganization by statute, since the power of the Legislature to reduce the number of departments, ". . . by consolidation or otherwise," is expressly recognized. Nevertheless any law designed to re-group two or more existing departments under a new name almost surely would be held unconstitutional as creating a new department. This gives to the executive establishment a rigidity unknown both to the national constitution and to those of a number of states. An administrative machine should be reasonably stable, it is true; but it should also remain reasonably flexible, and it is by no means certain that one attribute must (or should) be sacrified to the other. Whatever the attributes of those who revised the State's Constitution to prohibit statutory establishment of new executive departments, they were not so wise as to be able to foresee the administrative needs of the future. The prevailing concepts of public administration have changed drastically since 1938, as have the tasks performed by government; but there has been only one state department created since that date, and that, perforce, by constitutional amendment. Students generally agree that the revising fathers erred in writing New York's basic administrative structure into the Constitution.

On the firm base provided by the Constitution, the Legislature has elaborated the rambling structure described in Chapted III. The organization is not complicated by comparison with the national government's establishment in the water-resource field, but among the states New York stands unenviably near the top when judged by the criterion of complexity. If this basic finding be accepted, the reader may supply his own adjectives without much fear of successful refutation. The State's adminstrative structure for water-resource management is not only complex; it is also diffuse, fragmented, and largely without sense of general purpose or direction. The Governor might be expected to provide leadership for this splintered array of agencies, for by general repute he is a "strong" Governor; but his potential leadership of the water-resource agencies somehow fails to materialize. The familiar device of the *ex officio* board as an agency for coordination largely fails to achieve its purpose.

These are strong words, and in some circumstances might be dismissed as the complaint of a frustrated theorist. This customary escape hatch, however, was closed by a bill (No. 4351, Int. 3688) introduced into the State Senate on March 13, 1958. This bill, presented for study purposes on behalf of the Joint Legislative Committee on Natural Resources and the Temporary State Commission on Irrigation, corroborates the above testimony in a number of particulars. Section 1, titled Legislative Findings, reports (in subsection i) that:

> Over the years, in response to various water needs and water problems, a number of agencies have been given functions relating to water resources. There is need to coordinate more fully the present and future water resource activities of these various agencies.

The bill states further (in Section 1, subsection a) that:

> Long range, comprehensive planning for the conservation, development and beneficial utilization of water resources apparently appeared unnecessary [in the past], and has never been undertaken on a state-wide basis.

It refrains from stating outright that the present fractioned structure is incapable either of coordinated action or of comprehensive planning, but the import of its language is unmistakable.

The bill comes out unequivocally (in Section 1, subsection f) for ". . . the development of multiple-purpose water control and storage plans and programs . . . ," but passes over the evident fact that the patchwork organization makes the development of such plans and programs virtually impossible. Harking back to the case studies, it will be recalled that each treated of a particular problem and that none related even indirectly to multiple-purpose projects or to basin-wide planning. The principle championed by the study bill was honored in the breach, in these four instances at least, by a number of specialty agencies seeking to achieve individual, single-shot purposes. There was no concerted attack upon a broad problem because the legislative definitions of mission of the agencies involved do not admit of integrated planning or action in terms which are meaningful.

The divorce of the department head from operating responsibilities and the substitution therefor of interdepartmental boards in important areas (in water power and control and in pollution control, for example) has produced two grave consequences for administration. First, basic responsibility for decisions, failing to find strong executive leadership and eluding the clumsy grasp of the boards, has come to rest in the principal staff officers of the operating agencies. These are the top civil servants of the various agencies; they are experienced—several of them have served fifteen to twenty years; and they are generally competent within the fields of their specialty interests. They are, of course, thoroughly conversant with the rules and regulations of the civil service. More than this, they recognize a self-imposed, informal code of ethics under which each defers to the judgment of every other with respect to his particular area of activity and none asks embarrassing questions. The result is a comfortable atmosphere characterized by good manners, good will, mutual tolerance, and a general spirit of live-and-let-live.

Another and less agreeable way of saying the same thing is this: in respect of its water-resource agencies, at least, New York has developed a mature bureaucracy, with all that concept implies. There appears to be little or no sense of the administrative tension necessary to keep an organization on its toes. The rewards are not for initiative but for stability, not for daring but for sobriety, not for the hustle which leads to discomfort but for the quiescence which induces relaxation. The intellectual attitude is not one of inquiry into things new but rather one of satisfaction that old things go well.

It is difficult to pass this judgment without seeming to criticize individual officers. In truth, however, the emphasis here is not upon the civil servant but upon the civil service, or better upon the spirit of bureaucracy which is a constant threat to any long-established civil service. An experienced observer spoke of the "stagnation" which characterizes New York's upper civil service. That is too strong a word, yet it does indicate the nature of the problem. It does suggest the futility of looking to the top level of the civil service for a fresh and imaginative approach to the problems of water management in New York.

A second result of the absence of strong executive leadership is found in the tendency of the permanent administrative officers to seek close liaison with their opposite-number legislative committees. A similar tendency is found in Washington, where the U.S. Army Corps of Engineers, as a single illustration, enjoys extraordinary legislative support in part because of judicious cultivation of key committees, not excluding influential members of the Congress as well. Note has been made (in Chapter II) that a number of New York's programs found their origin in the active interest of this or that legislative committee. This is hardly surprising, in view of the fact that most of the State's water programs rest upon legislative authorization. What is somewhat surprising, and at the state level somewhat unusual as well, is the close continuing relationship which has grown up between some administrative agencies and the legislative committees which fathered them. The history and present status of the Water Pollution Control Board is a case in point. Established in 1949 in pursuance of a recommendation of the Special Committee on Pollution Abatement of the Joint Legislative Committee on Interstate Cooperation, the Board has attached itself to (or has been adopted by) the successor Joint Legislative Committee on Natural Resources (which was created in 1951 in part to take up the duties of the discontinued Special Committee on Pollution Abatement) in a way to suggest an agency-principal relationship. On the one hand, the Water Pollution Control Board as a matter of policy defers to the judgment of the Committee, while on the other the latter frequently takes action which in another context would be regarded as administrative. The relationship is perhaps best documented by the fact, noted earlier, that the Board's annual report is printed as a part of the Committee's yearly report.

It would be difficult to over-estimate the consequences of these administrative agency-legislative committee relationships (the one cited is not unique) for administration. It is quite true that New York's machinery for water-resources management is without head or guidance or central repository of responsibility, and that, as the natural resources committee has put it (in the study bill referred to above), "There is need to coordinate more fully the present and future water resources activities of these various agencies." It is also true that, considerations of fragmented organization apart, there is not likely to be much meaningful coordination until there is a sounder administrative-legislative relationship, *in terms appropriate to administrative considerations,* than that which now prevails.

Attention thus far has been focused upon state administrative machinery as such, that is, upon agencies with statewide jurisdiction. It is necessary now to take note of the *ad hoc* districts and agencies which are being created in increasing numbers for the pursuit of particular programs. Such units are established (or authorized) by the State, although they come into existence normally in response to local stimulus, frequently following presentation of a petition requesting action. Some of them receive substantial Federal financial assistance, and so operate in virtual fiscal independence. Financial autonomy often is translated into political autonomy, with neither State nor city exercising authority over them or assuming any significant responsibility for their activities. Fortunately not all local or special agencies authorized by law actually materialize as operating units, but enough do to contribute substantially to the disintegration of the State's water-management activities. The study bill mentioned above authorizes the establishment of "regional planning and development boards," in defiance of the principles of administrative coordination, comprehensive planning, and multiple-purpose basin development espoused earlier—and then places the "regional boards" at the mercy of the local board (or boards) of county supervisors. Not least among the unfortunate consequences of this kind of particularism is precisely the unplanned character of the developments which follow. As in the case of the more purely state agencies, the two major problems encountered here concern responsibility and concert of public purpose.

Paralleling the official organization is a network of private interest groups which exercise profound influence on public action. Representative of these groups are the numerous local conservation councils and alliances, fish and game clubs, sportsmens clubs, hiking clubs, bird study groups, trail hound associations, and wilderness societies. State organizations include the Conservation Council, the Adirondack Mountain Club, the State Association of Garden Clubs, the Forest Preserve Association, the Wilderness Society, and the New York Division of the Izaak Walton League, to name at random half a dozen of the more influential ones. Some of these have active programs, some come to life only when they sense a threat to their special interest. Normally their existence is placid enough; but they keep a jealous eye on developments in their areas of concern and are able to bring powerful forces to bear either in support of or in opposition to a proposed program. They maintain relations, sometimes active, more often dormant, with their opposite-number public agencies, which understand well and embrace the resulting tacit mutual aid agreements. The services rendered by these spokesmen of special interests are substantial; no less real are the obstacles they interpose to both the definition and the realization of public policy. They constitute, in effect, an unofficial government without whose support the legally responsible agencies almost always are hampered and sometimes are powerless to act. They contribute significantly to the atomization of government, complicating vastly the problem of administrative coordination and adding their bit to the cacophony which makes impossible anything approaching comprehensive planning.

A final consequence of New York's jungle of agencies in the water-resource field is found in what may be called the lack of public visibility. It is wholly unlikely that there is a public official in the State who does not pay lip service to the principles of democracy; yet democracy is not well served by a government which is so complicated in organization and so mysterious in operation as to be beyond citizen comprehension. The people cannot be expected to express intelligent interest in or take knowledgeable action on public issues which are obscured by underbrush. A *sine qua non* of democracy is a government where public visibility is high. New York's administrative machine is not calculated to admit the requisite amount of sunlight.

Summarizing, New York's structure for water-resources management is inflexible, complex, sprawling, headless, shadowy, and withal ill-organized for the task of planning and administering a total water-resources program for the State. In the absence of vigorous executive leadership, responsibility falls principally on the chief staff officers, who are permanent civil servants. With the passing of the years, these officers have settled down into a close and contented bureaucracy, which by nature tends to stifle initiative. Further for want of executive leadership, they have sought and found haven in a few continuing joint legislative committees, which have arrogated to themselves functions and powers normally considered to be administrative in nature. Numerous private interest-group organizations add measurably to the official confusion. The coordination and comprehensive planning correctly identified by the study bill as essentials of good administration clearly require attention of a kind and quality which thus far they have not received.

Part III: Conclusion

CHAPTER X

An Approach to Administrative Reorganization

THROUGHOUT this study have run as persistent undertones two re-current themes. The first has suggested the basic significance of administration for the success of a water-resources program. The second, by documenting the inadequacy of the present system of management, has indicated the need for administrative reorganization. It is to further development of the latter theme that we turn in this concluding chapter.

REFRAIN: FOUR MAJOR PROBLEMS

The study has brought to light a number of problems whose resolution is fundamental to the development of a satisfactory system of water-resources administration for New York. Among these, four have been singled out for summary emphasis here. The first concerns an appraisal of the State's water resources, to which three kinds of questions are relevant. First, what are the resources, where are they located and in what amounts, and what are the demands made upon them? Chapter I came to the conclusion that the basic data about New York's water are incomplete and in some part inaccurate as well. Second, what is the unrealized potential in each water program area, and what are the economic and social considerations involved in further development? Third, what problems stand most in need of attention? Systematic exploration of these questions would produce a standing record of the water resources that would prove invaluable in the continuing task of designing and managing a state water program. It is not suggested that these matters can be disposed of once and for all; on the contrary, the task of inventory and appraisal is a perpetual one, and emphasizes anew the importance of a permanent central agency dedicated to the management of water resources.

A second problem area has to do with the whole subject of state water-resource programs. Chapter II laid bare a maze of water programs, some of which go back more than a century-and-a-half, which has assumed its present form gradually over the years in response to specific needs and individual pressures for action. At this point in New York's history and at this time of stock-taking, a number of questions may legitimately be raised regarding these many-sided water programs. One such concerns the strength and value of each individual program, another the contribution made by each to the total program, another still the satisfaction with which all in the aggregate serve the gross program need. The end result of haphazard growth is a patchwork of programs which inescapably conveys the impression of unevenness in depth and coverage. The principal question therefore relates to the validity of the separatist approach when weighed against the total need. Here is a case in which the whole palpably is greater than the sum of its parts. The implications of this thought require careful and responsible consideration. An immediate reaction suggests the importance of over-all planning for program development.

On further reflection, attention turns to the whole process by which decisions are made in the field of water resources. Private businesses—power companies, paper mills, chemical plants, oil refineries, to name but a few—make decisions every day which have profound significance for the water resource. Many such decisions require approval by a public agency, even though private enterprise initiates the action. Also representing private interests are the innumerable organizations which concern themselves in such subjects as the Forest Preserve, water pollution, fishing streams, hunting grounds, hiking trails, parks, and water for agriculture. While these manifold organizations claim to voice the wishes of "the people," we must look to government for responsible public representation. There at the state level are found Governor and Legislature, often of different parties and so traveling different roads and frequently seeking different ends; the courts, whose decisions regarding water policy are basic; and no fewer than twenty-five administrative agencies, whose operational decisions give meaning to programs in practice. There are found, too, many Federal agencies with significant powers of decision, and in addition a multitude of local units which, though nominally subject to state supervision, often operate with almost complete independence. The net result of the multiform and uncoordinated actions taken by this army of decision-makers, public and private, is a babel of voices to which one listens in vain for a reassuring undertone of common purpose or policy.

The problem will be readily identified as that of achieving a reasonable definition of the public interest and devising means to give it responsible statement in programmatic terms. What do the people want of their water resources, and how can their wants and needs best be met? While the role of private interests will remain substantial, that of government must be judged paramount. A fundamental problem, then, is that of organizing government, first to achieve a balanced, inclusive program, second to make it effective through sound administration.

This leads to consideration, thirdly, of the role which government, in the present instance the State of New York, will elect to play with respect to water resources. For present purposes, we may confine the discussion to two aspects of the question. First, how broad a program should the State pursue in the regulation and development of its water resources? What should be the scope of its action with regard, for example, to hydroelectric power? New York has been in the water transportation business for over 140 years, but until recently there was virtually no public power in the State notwithstanding nature's generous endowment of hydroelectric sites. This situation will change drastically with completion of the St. Lawrence and Niagara projects, but even then sizable reserves of undeveloped hydro power will remain. What is the difference between operating a power plant and operating a canal? Is it that one produces a good return on investment while the other does not? And in any case, what is the reason for the striking difference in state policy between water transportation and water power? Who will develop the quite considerable power resources of the inland rivers? The relative merits of public and private development present a question of the first magnitude in pondering the role of the State in the water-resource field. It is a question which recurs in various program areas, though perhaps nowhere else in such dramatic form as in the domain of hydroelectric power.

Another dilemma which confronts the State in delineating its water-resource role rises from the intergovernmental character of the water problem. Bearing in mind the fact that the manifold aspects of water-resources administration are highly interrelated and call for concerted attention, what responsibility should be assumed by the various governments for what segments of the total problem? What would constitute a sound distribution of duties between the State and the Federal government in such areas as flood prevention and control, pollution abatement, municipal and industrial water supply, watershed management? In view of the growing emphasis on basinwide development, and further in view of the fact that most important basins cross state lines, what are and what should be the respective roles of Federal and state governments in river valley planning, development, and administration? The question has immediate relevance to the Delaware Valley, for which a general plan is in process of formulation even now under Federal leadership. There is room here for the kind of inventiveness that produced the federal system out of the chaos following the revolution.[1]

The interrelatedness of water problems finds further illustration in the part played by local governments in the management of water programs. The chief problem arising from local activities may be instanced by reference to the concept of multiple-purpose development of a facility, for which most local units lack either the resources or the interest. Can the State afford to condone the single-purpose exploitation of a site which holds promise of substantial multiple-purpose benefits? The question will be recognized as relevant to the recurrent issue which lies at the center of the problem of local participation in water programs. Involved are questions of the desirability of special-purpose districts, authorities, and agencies, and the powers and relationships of such units if they are to be established. There is a growing tendency in New York to make use of the special water districts which have long existed in profusion in some states. In this atmosphere, vigorous state supervision is the only effective antidote to a rampant particularism in which consistency in program is lost in a welter of local practices.

The fourth problem selected as worthy of mention here concerns New York's system of water-resource law, which was the subject of examination in Chapter IV. There the conclusion was reached that the riparian doctrine, embraced by the courts from the beginning of statehood, appears to leave something to be desired under the conditions of modern water-resource practice. It was not suggested there, nor is it implied here, that the law of riparian rights should be renounced in favor of its principal alternative, the western system of prior appropriation. At the same time, it must be recorded that a number of eastern states long devoted to the riparian system are weighing the possible advantage of a change, that one such actually has adopted the appropriation system, and that there is considerable dissatisfaction in New York with the water-resources law currently administered by the courts. A thoroughgoing study of New York's water law therefore would appear both useful and timely. A number of questions suggest

---

[1] See *River Basin Administration and the Delaware* (Syracuse: Syracuse University Press, 1960), for a detailed examination of this subject.

themselves for consideration. What is the nature of the reported dissatisfaction? Whence does it arise, and for what reasons? How far-reaching is it? How genuine is it? Are there problems arising from the law which do not find expression in vocal complaint? In particular, does the law place undue limitations on the development of the water resources? What remedies would appear adequate to meet the problems discovered? As noted earlier, there are steps that might be taken with respect to water-resource law both to alleviate the harshness of the riparian doctrine and to improve the system of judicial administration. What is required is not doctrinaire rejection of one system and acceptance of another, but a systematic examination of the problems encountered (or likely soon to be encountered) and action designed to resolve them.

Some Observations on Administrative Reorganization

Each of the problems examined thus far points in the same direction; each assists in setting the stage for consideration of the central problem of this study. What is needed above everything else, they say over and over, is a single administrative agency to represent the state authoritatively in its dealings with water resources—an agency that would build a comprehensive and cumulative record of water resources, that would fit individual activities together into a genuine water program, that would lead the way in delineating the role to be played by the State with respect to its water resources, that would speak with an authoritative voice for the State in its relations with other governments in the field, that would substitute visibility for the mystery which now prevails. The need insistently voiced for a central water agency highlights the core problem, that of administration. It has come to be understood almost universally of recent years that the success of a public program depends basically upon the effectiveness of the organization which administers it. New York's water problems, actual and potential, arise not from any absolute shortage of water but rather from lack of ingenuity and enterprise in managing the generous resources available. The problem is not one of limited supplies but of limited capacity to administer.

Even so, New York's need to reorganize its administrative structure, while pressing, is not imperative. Those who proclaim that "The time has come when we must set our governmental house in order" underestimate the resilience of democracy and the patience of the people; for it may not be truly said that administrative reorganization is at any given time, or is likely to become, an absolute necessity. There have been instances of complete breakdown of government, but they have been so few as to serve only to prove the rule that, in the absence of some catastrophe, government is capable of struggling along indefinitely under indifferent administration. And New York's water-resource machinery is by no means so bad as to pose the threat of collapse. Given normal operating conditions, it can continue for a long time to perform at a level which, if not outstanding, is at least good enough to get by. And so the State has an option: it can go along as in the past, playing the role of reluctant middleman and watching functions flow upward to the Federal government and downward to the local units; or it can assert a vigorous interest in the contemporary world and take action to halt the flight of functions from Albany. For the former course the existing system of management probably will continue to suffice; for the latter, reorganization of the administration is an essential first step.

The history of New York's administrative organization for water resources, detailed in Chapter III, is replete with attempts to bring order out of confusion. The creation of the Conservation Department climaxed one major effort. There have been other attempts to coordinate the activities of various agencies concerned with this or that segment of the field, some of them reasonably successful. For the most part, however, the measures taken have been half-hearted and half-way, and the conclusion is clear that their achievement in molding a symphony orchestra out of what remains essentially a group of individual performers has been quite modest. The problem is and for many years has been to establish an over-view, to the end that an integrated administrative structure may be created to bring order, purpose, and plan to the field of water-resource management.

There are, broadly speaking, two ways of approaching the problem of administrative reorganization. The first finds example in the Temporary State Commission on Coordination of State Activities (sometimes referred to as the "Little Hoover Commission"), a "temporary" body created in 1946 and continued since to study New York's departments and agencies and recommend legislation for the improvement of administration. This is the way of gradualism, which seeks to bring about the amelioration of ills by easy stages. In the spirit of gradualism, countless small moves may be made short of a frontal assault on the central problem. Administrative duties may be modified, responsibilities re-defined, relationships adjusted, programs elaborated or restricted. An occasional agency may be abolished, though rarely a very important one, and under extreme provocation (and in the fullness of time) two or more may be consolidated. Gradualism is the path along which both administrative agencies and programs travelled in arriving at their present places. It is the normal course of public action; it is the standard way of procedure of the Legislature, which is perpetually nibbling away at individual bits of a problem. If it does not hold the threat of spectacular failure with respect to governmental reorganization, neither does it offer the promise of dramatic achievement. There comes a time when the need for administrative reform is beyond its depth.

The second approach to administrative improvement calls for frank recognition of organizational shortcomings and forthright action to eliminate (or ameliorate) them through thoroughgoing administrative reorganization. The obstacles confronting deliberate and drastic reorganization are formidable indeed. They inhere initially in public apathy. Most New Yorkers assume that the State has plenty of water, and this psychology of abundance is a block in the road toward administrative improvement. "When the well is dry then we know the worth of water," Benjamin Franklin observed long ago. New York's well has not been dry, or has been dry so infrequently that the lessons of scarcity remain unlearned. Again, the State's water-resource agencies are believed by the people to have done, and to be doing, reasonably good jobs in their several programmatic fields, and there is therefore no widespread popular demand for reorganization. Yet again, reorganization is not an article easily sold to the people; for in the absence of official gross neglect, inefficiency, or scandal, it is lacking in popular appeal. Public indifference, then, or lack of public knowledge of or interest in so technical a subject, is the first obstacle which lies in the way of administrative reorganization.

Another arises from the possibility, perhaps even the likelihood, that significant reorganization would be found to require the creation of a new department, which would necessitate an amendment to the Constitution. The process of amendment is a complicated and deliberate one, calculated to ensure overwhelming legislative and popular support for a proposed amendment which in the end wins approval. It is not easy to visualize a situation in which an amendment calling for drastic administrative reorganization would command the support necessary for approval, excepting, perhaps, in the event of a movement for general constitutional revision. Some useful things could be achieved by legislative action, but the probable need to amend the Constitution cannot be dismissed. An amendment would be extremely difficult to bring about.

In the next place, the forces opposing reorganization ordinarily are stronger than those favoring. Such forces include history and tradition, which support the *status quo;* cliental groups, who "know the ropes" (and the officials) and therefore mistrust change; interest organizations with hobbies to ride and skill at riding them in familiar surroundings; and the water agencies themselves, whose staffs, convinced that they are doing a satisfactory job and often oblivious of such lateral issues as program coverage and relations, react defensively, sometimes violently, to suggestions for reform. Not to be omitted from this listing is the Legislature, which through several of its joint standing committees maintains jealous relations with the water-resources agencies. Now and again a favorable combination of circumstances, as the simultaneous election of a dynamic Governor and a sympathetic Legislature, a far-reaching public scandal involving state officials, or an emergency, will set the stage for a state reorganization. In the absence of some such dramatic occurrence, the odds strongly favor those who support business as usual.

Still another factor which complicates the problem of administrative reorganization is implicit in New York's political behavior. Its nature is suggested by a simple but pregnant historical fact: since 1910 the Democrats have controlled the governorship thirty years, the Republicans eighteen, while the Republicans have controlled the Legislature forty-four years, the Democrats but four. Governor and Legislature thus have been of different parties over half the time during the last almost fifty years. The significance of this fact becomes clear when it is remembered that any important change in governmental structure requires collaborative action by the two branches. The political hurdle in the path of drastic administrative change therefore is a high one over half the time, although New York's experience indicates that it can be taken. And there is the fact that, since 1910, Governor and Legislature have been of the *same* party almost half the time.

A final barrier is of a technical character. It arises from the nature of the water resource, which is more diffuse and less susceptible of unified treatment than some other program areas. Water is intimately related to a number of long-established programs, among them navigation, health, agriculture, forestry, and recreation. Moreover the relationship is in no sense an artificial one, but on the contrary stems from legitimate if sometimes secondary program interests. Those who would reorganize the State's water-management machinery therefore face a difficult dilemma. On the one hand, water activities as such probably can be best administered in a separate water agency; but on the other, such activities are naturally related to other programs which would be adversely affected by their transfer. This dilemma is in no wise limited to the area of water-resource administration, though it appears in striking relief there. It poses a technical problem of the order of the Gordian knot.

The presence of these (and other like) obstacles neither minimizes the shortcomings of the present administrative structure nor obviates the necessity for its thorough reorganization. The officially identified need for coordination of administrative agencies and for long-range, comprehensive planning, and, by clear implication, for the administrative reorganization necessary to make possible achievement of those ends, is clear and present. Administrative structure is not everything, to be sure; indeed some writers on public administration for two decades have downgraded organization, seeking to substitute the pragmatic test of whether the assigned job gets done. Without succumbing to the temptation to set up a smooth, symmetrical organization as an end in itself, it may be argued nevertheless that a logical structure is more conducive to effective performance than one scattered all over the administrative lot. The ultimate test, in administration as in baking, is found in the edibility of the product; but the tools of the workman, it may be surmised, are of fundamental importance in both callings. Organization is a basic tool of administration.

Once the decision to proceed with reorganization has been reached, the alternative organizational forms available are clear enough. In essence, they are two, although there are several possible variations from each. First, a unified department of water resources can be established and responsibility for the major water programs concentrated there. California has taken this route, though it is doubtful that an equally strong case for a separate department can be made in water-rich New York. Second, a water resources division or bureau with jurisdiction over the principal water activities can be created, or an existing one expanded, in one of the major departments. This seems the better of the two options, and the Conservation Department appears to be the logical administrative locus of a water agency with broad powers. The fundamental problem at this point, however, is not the form to be assumed by the new organization—the form will emerge in due course, and in any event its consideration in detail at this time would be premature. The basic problem concerns the leadership of the reorganization movement and the procedure necessary to its successful issue.

Five major centers of interest in the problem of water-resource administration can be identified. The first is the public at large, whose concern is genuine but as a rule amorphous. Only on rare occasions does a popular figure emerge to galvanize the public into action with respect to a routine peacetime problem. It is next to impossible to imagine the emergence of such a one to organize and lead a popular movement for administrative reform. Another center is found in the manifold special groups which have interests in this or that particular water-resource area. There is considerable leadership latent in these groups, but it is hardly to be expected that it will spring to action in behalf of general administrative reorganization of the water-resource agencies. Such groups are almost uniformly limited in interest and therefore narrow in outlook, and so, not surprisingly, are not known for the breadth or variety of their concerns. They offer little promise of help in our quest for leadership.

Yet another major center of interest is found in the administrative agencies themselves. Their involvement is, of course, deep and genuine; further, they are frequently ably staffed. That there is dedicated leadership among the agencies is not to be doubted. It is, however, for the most part technical leadership, more preoccupied with operational considerations than with administrative organization as such. One would not expect the professional staffs to provide spark or guidance to a movement for administrative reform.

Next there is the Legislature, among whose members are several who have devoted years to the problems of New York's water resources and their management. Note has been made of the tie-in between certain water-oriented committees and the apposite administrative agencies. These close and in some instances long-standing relations have resulted sometimes in emotional ties to programs which cloud the broad view so necessary to the task at hand. The Legislature may be looked to for widespread interest and for considerable information and experience regarding special water-resource agencies; it cannot be depended upon for the initiative necessary to set in motion a movement for administrative reorganization.

This leaves, as the last and most important center of interest in New York's water-management problems, the executive department, and more particularly the Governor himself. It is precisely the responsibility of the Governor to do what no other locus of interest is fitted to do, namely to see a problem whole and in perspective, and to bring vigorous action to bear in its solution. The Governor will not wish to move without the knowledge, indeed without the active support, of all parties at interest; on the contrary, he will wish to involve every interested organization and agency at the earliest possible stage. This is true particularly of the Legislature, without whose positive support he will not go far. But he and he alone occupies a place of leadership with respect to all the people. His position makes him the State's chief administrative officer. As such, his obligations are of course many and varied; but in the present context they center on his responsibility for leadership in exploring the problems of the organization and reorganization of New York's governmental machinery. As of a particular moment he may have concluded that the State's whole administrative structure is in want of reorganization. Those whose primary concern is water resources should welcome an executive decision to join the issue of administrative reform on a broad front. The case made here, however, is a less ambitious one, namely that the machinery for water-resource management is in need of a thorough overhauling. This, too, poses a responsibility worthy of the hand of the Governor.

A Proposed Program of Action

Everywhere there is evidence of a growing concern on the part of the states for their water resources. The Council of State Governments reports that "Within the past three or four years, legislatures in more than half the states have created special water study commissions . . ." [2] Several states, passing through or shortcutting the study stage, have enacted legislation overhauling their machinery for water-resource administration. California created a Department of Water Resources with wide powers in 1956. Florida appointed a Water Resources Study Commission, which made its report to the Governor and the Legislature at the end of 1956. Oklahoma named a State Water Study Committee in 1955; the report of the Committee led in 1957 to an act setting up a Water Resources Board with fairly broad administrative duties. Closer home, most of New York's near neighbors have taken steps looking to the improvement of their water-resource management. To the east, Massachusetts created a Water Resources Commission with general coordinative power in 1956; and Connecticut, following a report

[2] The Council of State Governments, *State Administration of Water Resources* (Chicago, 1957), p. 59. Chapter 5, "Recent Reorganization Proposals and Legislation," deals directly with the subject at hand, although the whole of the study is relevant.

by a study commission, established a Water Resources Commission with broad planning and coordinative powers in 1957. To the south, the New Jersey Commissioner of Conservation and Economic Development established a Water Resources Advisory Committee, which issued its first report in April of 1957. On April 5 and 6, 1957, a Pennsylvania Water Resources Conference met at the call of Governor George M. Leader. The Conference recommended

> . . . the establishment under proper authority of an advisory body with appropriate funds and powers to study and report, within a limited time, not to exceed two years, on the administrative organization and legislative requirements which may be necessary or desirable to initiate the preparation of a long-term, comprehensive plan for the effective control, development, conservation and utilization of the water and related land resources of the Commonwealth.[3]

Here is a preparatory body which contemplates long-range action in process of gestation.

These agencies exhibit wide variety. Some are called commissions, some boards, some committees; some are designated advisory bodies and some study groups, while some serve without qualifying adjectives; most rest upon statutory authorization, but a few result from administrative directives; some include lay citizens alone and some officials alone, but most contain both; some are financed by legislative appropriation and some by private contributions, while some have no visible means of support; missions range from the sky's-the-limit to grumpy adjurations to find ways to provide more service for less money. And some emerge as full-blown administrative units, with or without benefit of antecedent study commissions. Precedent therefore may be found for any course one may choose to recommend.

Initial action here proposed for New York is simple and direct: it is that the State join in the parade by creating an exploratory Water Resources Commission. In view of the responsibility for leadership earlier attributed to him, the Governor might proceed forthwith to draft a plan of administrative reform and lay it before the Legislature for approval. Such a course, however, would appear to be ill-advised. For one thing, the Governor will need technical assistance on any plan he may choose to sponsor; the proposed Commission would provide him with such assistance. For another, the period required for the work

---

[3] The Governor's Water Resources Conference, *Report to Honorable George Leader, Governor of the Commonwealth* (Harrisburg, Pennsylvania, May 6, 1957), p. 3.

of the Commission could be used for public education, with the Commission itself playing a prominent part in that process. For another still, the Commission could be employed to enlist the cooperation of many groups and agencies, foremost among these the Legislature itself. Finally, the appointment of such a body would seem entirely normal to New Yorkers, who are well acquainted with the study commission device, seen most recently in the Temporary State Commission on the Constitutional Convention.

The Commission should be an official body, and should therefore be authorized by law. To govern the composition of the Commission, four principles are proposed. First, *ex officio* memberships should be avoided. There are several reasons for this: the top officials in the major water-resource agencies are already overtaxed by such memberships; there are so many important water-resource agencies that the Commission would be over-balanced with *ex officio* members if recognition were given to all, while it would be difficult to choose from among them a small and representative number; and New York's many *ex officio* boards and commissions have not achieved outstanding success in the past. The goal to be sought is not official but public representation.

Second, interest-group representation should be eschewed in designing the Commission. The special pleaders—forever-wilders, farmers, shippers, rod-and-gun clubbers, dam builders, city dwellers, doctors, industrialists—must assume a good share of the responsibility for the present disorganized state of New York's water-management program. Once more is the whole greater than the sum of its parts, for the particularized interests of the specialty groups do not aggregate to the public interest.

A third principle, which flows from the first two, is that the Commission should comprise a body of citizens who have demonstrated their interest in public affairs through significant lay (uncompensated) public service. The Commission must possess two attributes above all others: it must clearly have the stature to rise above partisan, professional, and special-interest involvement, and it must enjoy the respect and confidence of the public in the highest degree. The enterprise will profit from the concentration of public interest and attention on the work of the Commission, which suggests that membership in that body be held to a small number.

The fourth principle arises from the basic fact that the Commission is to be a public, an official, body. It is to be an agent of the State, and as such must be chosen in a way to ensure its public character. It is proposed that the Commission be limited to seven members, two to be appointed by the Temporary President of the Senate and two by the Speaker of the Assembly, and three to be named by the Governor. It will be observed at once that this proposal seeks to involve both executive and legislative branches through representation on the Commission. It is important that both Governor and Legislature participate in the enterprise from the beginning, since the two will have to agree if any significant action is to be taken in the end. At the same time— and this is an important point—the legislative representatives need not be members of the Legislature; on the contrary, the Temporary President and the Speaker may designate as their four members any individuals they feel will add strength to the Commission. For the other three appointees, it will be incumbent upon the Governor to search the State for outstanding citizens. Upon these seven men, upon their integrity and their stature and upon the public respect and confidence they command, will depend the success of the Water Resources Commission.

The mission of the body can be deduced from what has gone before. In general terms, it will be directed to make a study of the state administration of the water resources of New York. More specifically, it will be instructed to

1. make an inventory of the State's water resources and an evaluation of their developmental possibilities;

2. delineate the role of the state government with respect to the water resources;

3. re-define and elaborate a state water program (or series of programs) in light of the objectives set;

4. evaluate the prevailing system of water resources law in view of new conditions and changing demands; and

5. essay the adequacy of the existing administrative machinery and procedures to achieve the goals set. Be it noted specifically that the Commission's responsibilities are not limited to consideration of administrative structure, which nevertheless is regarded as central to its mission.

To these ends, the Commission will be instructed to make recommendations, both general and particular, for action in every field covered. It will be directed to criticize where criticism is in order, to propose remedial steps where existing organization or procedure is found unsatisfactory, and to recommend new measures where gaps are discovered. It will be instructed finally, and above all, to fix its eyes on the distant horizon and plan for a comprehensive water-resource program for the State, and for an administrative organization capable of managing such a program.

In the pursuit of its duties the Commission will need to examine not only New York experience and practice but relevant experience elsewhere. Other states, where the pinch for water has been felt longer and more sharply, have much to teach New York about water-resource management. As sources of information, the Commission naturally will turn first to the State's own agencies, legislative as well as administrative, which it will discover have done a great deal of work in the field. The heads of these bodies will be directed by statute to cooperate with the Commission in every possible way. Second, the Commission may wish to conduct public hearings around the State, in the manner which some other agencies have found useful. When all ready-made and convenient sources of information have been drained, however, the Commission will discover that many important subjects remain about which it knows relatively little. This will be true not because existing agencies will not have a mass of data on individual programs, but because the commission will want information in part of a kind that no special-program agency can be expected to have. Third, therefore, the Commission will need a research staff of its own.

The staff of the Commission should be headed by a highly competent, widely experienced, well-paid executive director. What he will require in the way of research associates cannot be foretold with assurance, but certainly he will need one senior person for each substantive research area. Thus a professional staff of five or six experienced persons will be required. The time needed for such a survey as that contemplated is a highly variable factor, depending mainly on the scope and depth of the studies undertaken. A substantial job can be done in two years, and that is the time limit suggested for the project. To minimize the all too familiar temptation to extend the life of the Commission, the limit should be as firmly fixed as it can be made by law. Budgetary needs, like staff requirements, cannot be forecast with certainty, though it is clear that the survey contemplated cannot be done well without substantial financial support. In view of the task outlined, the figure of $250,000 appears not unreasonable.

Here, then, is the essence of the proposed program of action: creation of an official Water Resources Commission charged with the responsibility of surveying state administration of the water resources of New York, and support of the Commission at a level which will permit it to bring in a report two years from the date of its appointment. Such a step will by no means guarantee the immediate and permanent solution of New York's water resources problems. Such a solution, indeed, is visionary and not to be expected. It will, however, lead to recommendations which, if adopted, will provide both a base and an organization for making sound decisions regarding New York's water resources. And if this seems but a modest step forward, let it be remembered that in normal circumstances deliberateness is of the essence of administrative change.

Index